Journalism and Media Convergence

Media Convergence/ Medienkonvergenz

Edited on behalf of the
Research Unit Media Convergence of
Johannes Gutenberg-University Mainz (JGU) by
Stefan Aufenanger, Dieter Dörr, Stephan Füssel,
Oliver Quiring and Karl Renner

Herausgegeben im Auftrag des
Forschungsschwerpunkts Medienkonvergenz der
Johannes Gutenberg-Universität Mainz (JGU) von
Stefan Aufenanger, Dieter Dörr, Stephan Füssel,
Oliver Quiring und Karl Renner

Volume/Band 5

Journalism and Media Convergence

Edited by
Heinz-Werner Nienstedt, Stephan Russ-Mohl
and Bartosz Wilczek

DE GRUYTER

ISBN 978-3-11-048456-4
e-ISBN 978-3-11-030289-9
ISSN 2194-0150

Library of Congress Cataloging-in-Publication Data
A CIP catalog record for this book has been applied for at the Library of Congress.

Bibliographic information published by the Deutsche Nationalbibliothek
The Deutsche Nationalbibliothek lists this publication in the Deutsche Nationalbibliografie;
detailed bibliographic data are available in the Internet at http://dnb.dnb.de.

© 2013 Walter de Gruyter GmbH, Berlin/Boston
Typesetting: PTP-Berlin Protago-TEX-Production GmbH, Berlin
Printing: Hubert & Co. GmbH & Co. KG, Göttingen
♾ Printed on acid-free paper
Printed in Germany

www.degruyter.com

Contents

Part 3: **PR, Journalism, and Convergence**

Part 4: **Search Engines and Social Media**

Part 5: **Conclusions**

Part 1: **Quality Journalism under Pressure**

Stephan Russ-Mohl, Heinz-Werner Nienstedt, Bartosz Wilczek
Journalism and media convergence

An introduction

---

1 Problem diagnosis from different perspectives

Journalism is under increasing pressure, due in large part to the phenomenon of media convergence. Not only does media convergence redefine the tasks of journalists and newsrooms, it also reshapes the business environments of media companies.

In this book, international media practitioners and researchers describe and analyze the relationships between media convergence and advertising, public relations, social media and other areas of communication posing challenges to journalism. Concurrently, they contribute to and participate in the search for new, innovative ways to sustain good journalism.

This book documents contributions to a workshop at the Research Unit Media Convergence of the University of Mainz. It differed from many other conferences due to its dramaturgy: We tried to bring media practitioners and researchers together, and hoped that expertise from different fields which are closely related to journalism would result in more accurate problem diagnosis, if not problem-solving. Initially, the intention was to begin each panel with an expert-practitioner in the specific field, and to thereafter offer commentary from a researcher and a journalist. However, we had to compromise here. This did not change our intention to gather ideas under a "rescue umbrella" to help maintain one of the most important services democracies require: a free and competent press working to keep watch over the powerful.

As most of us are well aware, the most admirable form of journalism has become an endangered species. Its traditional forms of print, TV and radio are converging on the Internet. If we don't care, and if we don't find new business models or new ways to garner support, high quality journalism may vanish into the Bermuda triangle of cyberspace. Journalism could become increasingly irrelevant for three major reasons:

- Search engines and social network sites offer new opportunities for advertisers to reach specific target groups more effectively.
- PR is increasingly instrumentalizing journalism and thereby challenging journalism's credibility.

- The public is unwilling to pay for online content, which may also result from journalism's increasing reliance on prefabricated PR feedings, and thus, the decreasing credibility of journalism.

In this book, practitioner expertise and research results have been combined with the ultimate goal of achieving a better understanding of how journalism may yet flourish under conditions brought about by media convergence.

Initially, Robert G. Picard, media economist and director of research at the Reuters Institute for the Study of Journalism at the University of Oxford, argues that there will always be a need for news and journalism, although the ways journalistic news has been financed, distributed and consumed are being altered profoundly. Moreover, the pace of change is affecting news media differently, and some countries are affected more rapidly than others. Therefore, the big question is how to finance and organize journalism in the 21st century – an issue which will be addressed in more detail in several contributions to this book (see articles from Browne, Gellenbeck, Meinhold, Nienstedt/Lis). In order to survive, news organizations must innovate and adapt to developments of the converging media world. Change is no longer a choice, claims Picard, "it is a requirement for news organizations!"

Thereafter Gabriele Siegert, a media economics professor and advertising expert from the University of Zurich, summarizes changes in the ad industry from a researcher's perspective, addressing how these changes affect quality journalism. Her contribution relates to two statements from renowned advertising practitioners, Sebastian Turner and Jens Erichsen, who also participated in the conference. Unfortunately, we were not able to document their contributions in full length. However, to provide the necessary context, we have condensed their original contributions here.

2 Significant conflicts and emotional eruptions

Sebastian Turner has not only been one of the leading figures behind the international advertising agency Scholz & Friends, he is also a co-founder of the German *Medium Magazin*, one of the two prominent trade journals for journalists in the country. Presently (as this book goes into print) he is campaigning for the position of mayor in Stuttgart. In his statement in Mainz (Turner 2011) he emphasized, "after centuries of mutually beneficial cohabitation, the relationship between newspapers and ads is nearing its end." Classified advertising was first to leave the joint household, retailer advertising followed, and according to Turner, "thereafter little is left." He anticipates "explosions, landslides and the like" as the divorce of editorial and advertising forges ahead with significant conflicts and emotional eruptions.

Turner foresees the largest ad sector vanishing. The shift toward online media reduces the reach of households ("Haushaltsabdeckung") of print media to beneath what is considered a critical mass. According to him, this forces the most important advertisers, the retail chains with many outlets, to search for other advertising channels as they require a high penetration of households. From his viewpoint, the new channel may not automatically be digital advertising. Flyers in every mailbox may be an even more compelling option – a service even newspaper printers and distributors can provide – though the editorial staff will no longer be essential for this service.

According to Turner (2011), the most crisis-resistant advertising sector "has already found a new home," pointing towards search engines and social network sites:

> The availability of cheap, highly-targeted advertising lures away the segment of the advertising market closest to sales – the segment most reliable even in a period of economic downturn. The person who is about to buy something (as opposed to someone who is not yet considering a specific purchase) makes himself identifiable by his search words ("convertible," "piano lessons," "plumbing in Eberswalde county") in the very moment he prepares his expenditure. The cost per interest and cost per order is collapsing. (Turner 2011)

This enables the new business models of advertisers to thrive without print media advertising, according to Turner.

Turner (2011) also pointed to the fact that news organizations have not recognized that the revolutionary new changes in business models are rarely spearheaded by traditional market leaders. "Don't expect the sailboat manufacturers to be first in the motor boat business," he warned, providing proof in his own field. "None of the new mechanisms in advertising, such as the auctioning of ad space, were invented or even adopted early on by the dominating advertising carriers, i.e. newspapers and magazines." Turner stated that the new rules were established by players such as Amazon, AOL, Apple, Craigslist, eBay, Facebook, Google, Microsoft and Netscape. "When news organizations entered the stage, they usually wrecked it," he said, referencing the merger of AOL and Time Warner and Rupert Murdoch's acquisition of MySpace to illustrate the underlying structure. Market leaders "forget to differentiate between the essential need of their customers and the habits developed due to technological needs that may change as technology changes."

Under these conditions, according to Turner (2011), publishing houses as "transaction cost owners" are under pressure. News organizations with giant resources like printing plants, distribution centers, truck fleets and armies of delivery people "are challenged by individuals who own nothing but a computer and a desk in their bedroom." As soon as "the quality provided by the small independent start-ups is perceived as being a peer to the traditional news companies, the old news companies will be eaten up by fixed costs."

Turner sees local and regional newspapers not only losing dramatic income from advertisers, but also under attack from two angles journalistically. "National and international reporting of high quality is drawn from national media and specialists, and the local reporting will be challenged by micro local media."

Turner predicts that all-in-one-media offerings will be replaced by individually arranged news baskets: "The work of new, highly-respected individual authors will claim relevance, being rearranged by the audience into new individually composed general interest media. The general interest news providers of the past, the newspapers, will follow their soul-mates in the magazine business and disappear – such as *Life* in the U.S., *Quick* and *Neue Revue* in Germany."

Turner (2011) also addressed a development visible in the U.S. for the last years, suggesting journalists who have been laid off "will challenge the local papers with micro-local journalism." As the traditional local newspapers continue to lose circulation revenue and advertising income, they will have to reduce their editorial staffs. "This will not only reduce the quality of reporting, the unemployed journalists will start micro-local news offerings on the Web, establishing a completely new media category, micro-local media serving small neighborhoods in big cities and small villages in large districts," said Turner.

3 More complexity and more opportunity

Jens Erichsen, managing director of Carat Deutschland, the German branch of the world's leading independent media planning and buying specialist Aegis Media, contributed further insight about the development of advertising and the "divorce" of advertising from journalism. According to Erichsen, there is "more complexity and more opportunity" due to digital technology. "Around the globe there is a shift from mass media to masses of media, from mass audiences to masses of audiences, from broadcaster control to consumer control," he said (Erichsen 2011), emphasizing the new role of consumers. "They are embracing the future. They still love to be entertained and informed but now use media to interact and transact, to socialize and create – happily converging media to suit themselves."

He and his company identified three major "game changers" in advertising campaigns and brand communications. First, he suggested that "the days of 360-degree integration, with their campaigns of 'matching luggage' are behind us." Instead, "we need to increasingly think of how we can optimize a brand's continuous activity over 365 days per year. Few advertisers have the luxury of being 'on-air' every day of the year, so our task is to create programs of activity that allow the customer to engage at a time of their choosing." According to Erichsen (2011), this requires continuous listening and continuous iterations and approaches to planning and how content is programmed. "In fact," he said, "editorial thinking becomes crucial for brands. Successful brands break out of the fancy ad world and become generous, useful and personalized to connect with like-minded people around shared interests and causes."

Second, according to Erichsen (2011), "*creating time* becomes more important than just buying time." He believes in "planning for and creating situations that people will actively choose to spend time with us, where ideas can propagate actively and freely over the full extent of bought, owned and earned media. We believe in sustaining ideas that are born to live, not born to die. Hermetically sealed ad campaigns belong to the past, on-going projects generate sustained engagement." Thus, the creative advertiser even replaces the journalist in part as an entertaining storyteller.

Third, "whilst big ideas made sense in a broadcast world when media was scarce and controlled, in today's fragmented and increasingly social world only irresistible ideas will do," said Erichsen (2011). He added that irresistible ideas demand a response, they invite a next step, and hopefully take on a life of their own. "Becoming self-financing and self-perpetuating, they are more likely to get picked up and passed on, and they make it inevitable that fans will choose to get involved. Ideas which are remarkable, that get talked about, that are irresistible – these are the most powerful tools at our disposal to change behavior, culture and business."

According to Erichsen, *content, sociability, mobility,* and *commerce* are converging and driving convergence. "To exclude these elements from any communication program would be like leaving out the water when baking bread":

- – *Content* by far offers the most significant opportunities and the most substantial threats for journalistic media. Content is the main driver for any kind of conversation, and conversation is what brands are looking for. Conversation is the most effective way of communication. Research shows that conversation and recommendation are more effective than any other communication channel. So it is no surprise that brands try to seize new media to stimulate conversation. The ultimate way to seize content for a brand is to become a publisher. And indeed, we observed plenty of best practice cases like the Walkers Crisps sandwich campaign or Gatorade Replay where brands tried to engage consumers with their own content. Most brands engage on the level of entertainment, but services like "severe weather warnings from ERGO insurance" or "digital concert hall of Deutsche Bank" show, that brands increasingly compete directly with traditional media.

- Media usage becomes more and more *social*. Research from the U.S. indicates that 80 percent of mobile users use their mobile device and TV at the same time. In particular, TV highlights of major sports events multiply Twitter and Facebook usage. This harms a major function of journalistic media: the commenting and judging. In fact, the social aspects of media consumption make journalistic services – at least on big events – partly redundant.

- *Mobility* is expected to advance as a key vehicle for enhanced socialization. We are already seeing prescriptive social software apps like Foursquare that utilize the networks and friendships of users to drive mobile behavior. It is impossible to be out of touch or out of reach. Recent research shows that mobile devices are used not only out of home, but to the same extent in-house as well. The personalization of authority becomes an even broader trend throughout the mobile universe. The trend refers to pro-active and self-confident behavior selection on the part of technologically advanced consumers. Consumers can find and share intelligence in real time at any point in the day. This intensifies their power over the brand-owners but opens new business opportunities to journalistic media like location-based news or social news enhancement.

- Convergence enables social *commerce*. Social media and the technique of behavioral targeting make reaching the most valuable consumers far more precise than any journalistic media. Brands involve their consumers in the product development, product marketing and product services. Prominent examples are the crowdsourced development of the "Fiat Mio" in Brazil, the "Adidas MiCoach" in the U.S. or "Tchibo ideas" in Germany. These brand communities supersede the need for mass communication to a certain extent. Not all consumers will ultimately engage with a brand, around one to two percent of consumers engage actively. However, due to word of mouth and recommendations, the coverage of social media becomes increasingly interesting for brands. (Erichsen 2011)

Finally, Erichsen reminded us that "media are brands too." They can "actually learn how to position themselves in a converging world from brands. Engagement between media and audience is currently rather low and needs to be increased. New devices affect media consumption and offer great opportunities to adapt to new audience habits. Technology offers opportunities as well, for example the direct measures of what users seek and what they read" (Erichsen 2011).

In closing, Erichsen reflected on *Time* magazine founder Henry R. Luce, who believed that the primary duty of a journalist is to serve readers – a view worth remembering, though it only makes sense if readers have responsibilities to journalists too, and are at least paying for part of their earnings.

4 Herding behavior and social proof

The danger of collapse for ad-financed journalism has lead to those four book contributions, which directly discuss the future financing of journalism. Not only has this topic gained increasing attention from media experts and researchers (German media journalist Harald Staun (2012) is concerned about an overabundance of effort to rescue quality journalism), it is also an area where herd behavior of media researchers and media practitioners can be studied at large. Only a few years ago, there was a "unisono credo" on all major media channels distributing conventional wisdom: "Information wants to be free." As it had been free on the Internet for several years – and as zero cost is a very special price stimulating predictably irrational behavior (Ariely 2008) – there would be never ever again a chance to introduce paywalls.

However, a free, accountable press is an independent press, and independence is largely based on the financial soundness and well-being of the media industry. Looking backwards, it is also remarkable to note how the media and journalists undermined media independence and financial soundness with the way they've reported on the crisis.

The very word *paywall* is a dubious creation. Is the baker who charges for the donuts he sells constructing a "paywall" between him and his customers? By circulating the catchword together with the seductive, but nevertheless stupid formula "information wants to be free," journalists not only contributed to the herd behavior of others, but became victims of it themselves. If everyone believes in God or UFOs, this is unfortunately not yet empirical proof that everyone is right. However, it is evident that the believers become victims of a phenomenon which researchers call "social proof" (Asch 1951; Cialdini 1998) – a phenomenon which describes how easily we tend to believe what others, particularly the media, believe, and which explains why few trendsetters are able to prescribe new diets, new sports, new clothing, or even how a religious sect leader might successfully talk his followers into collective suicide (Dobelli 2011: 17). In a similar way, a few new media gurus were able to convince everyone that paywalls will never work.

Only three years ago, very few publishing houses – among them America's cheerleader, *The New York Times* – seriously reconsidered this notion that excellent journalism has a value and should therefore not be accessible for everyone at zero cost. Perhaps fifty years from now, when historians will write about the downturn of the U.S. newspaper industry they'll find that *The New York Times* prevented the industry from collapse by courageously implementing its metered paywall. At least *The New York Times* was already preparing the turn of the turn – to move from the so-called *Times Select* to free access and from there back to its metered paywall. This was accompanied by remarkable communication efforts, including intensive coverage of the newspaper industry's downturn, frequent presence of top editors like Bill Keller and Jonathan Landman in blogs, in chatrooms, and on podiums (for more detailed documentation, see Russ-Mohl 2009: 58 ff.).

5 The future of financing journalism

The four contributions dealing with future financing look at the problem from very different angles.

Heinz-Werner Nienstedt, media management professor at the University of Mainz, and Bettina Lis, assistant professor at the Chair for Media Management at the University of Mainz, argue that the print part of the newspaper business is still quite healthy in Germany when compared to other parts of the world. Editorial resources, which have not been significantly reduced in terms of industry-wide headcounts since the 1990s, can still be financed by the print part of the business. Online publications of newspapers, which are still free for the user with minor exceptions, may break even or have small profits or losses, but their profit and loss statements reflect only marginal online costs. Editorial costs are overwhelmingly allocated to print. Given the transition to digital consumption of media, online will have to bear parts or all the editorial fixed costs in the mid or long term. There are reasonable doubts about whether online advertising can fill this gap. Thus, newspapers have to turn to paid content models. Nienstedt and Lis discuss main obstacles for this undertaking as well as factors which indicate a success for such strategies on a mid-term time horizon.

With regard to keeping in close touch with readers, the German cheerleader among daily newspapers is most probably the leftist-alternative newspaper *tageszeitung (taz)* in Berlin. It is owned by a cooperative of readers and other activists. Konny Gellenbeck, longtime director of this cooperative, shares her experience in the so-far-successful fight for survival by community building and by an admirable communication strategy. Other quality newspapers still might learn a lot from it, though not every step taken by a niche project like the *taz* may be feasible for much larger quality newspapers like *Frankfurter Allgemeine Zeitung, Neue Zürcher Zeitung*, or the *Guardian*.

Karl-Heinz Ruch, who has been working for the *taz* since 1979 and who would in a more "capitalistic" enterprise be called the CEO, has explained that the business model of the *taz* will work only in affluent societies – though it is strongly based on the principle of solidarity (Ruch 2012). We agree: *Moral markets* flourish better if market participants don't have to fight for physical survival in their daily lives. However, the practical example of the *taz* (and similar efforts of Max Havelaar or the Tea Campaign – not to be confused with the Tea Party Movement – to get fair prices for agricultural products from the Third World) show that higher prices for certain products and services may be accepted as "fair" by a significant share of customers/audiences – if transparency is provided, and if the respective need to protect income is communicated actively. This has also been confirmed in experiments conducted by behavioral economists (Kahneman/Knetsch/Thaler 2000: 323 ff.).

Harry Browne, a lecturer of journalism at the Dublin Institute of Technology, takes an unconventional closer look at foundation-funded journalism, another funding model circulating with idealistic expectations, particularly in the U.S. Though there is no doubt that foundations can and should play an important role in stimulating changes and innovation in the overall society as well as in media and journalism, Browne is concerned about journalistic independence and points out that there are some risks involved in such funding – providing evidence with statistics as well as enlightening examples. The recent one million dollars grant from the Ford Foundation to the *Los Angeles Times* (Rainey 2012) might be added to his list. Even such a generous gift will only slightly alter the survival conditions of one of America's most prestigious newspapers. It certainly cannot compensate for the loss of staff, institutional memory, and quality caused by the newspaper crisis and – in this particular case – a fraudulent bankruptcy initiated by Sam Zell, the paper's most dubious investor. Due to specifications accompanying the grant, there will be a latent, invisible external influence on newsroom decisions concerning the issues to be covered in the newspaper.

Joachim Meinhold, CEO of the *Saarbrücker Zeitung Publishing Group*, argues that the "crisis rhetoric" about newspapers is highly questionable. He warns not to give up a successful business model and proposes to better optimize it. The traditional financing model for journalistic content enhanced by new but related businesses still enables a return on sales well into double-digit figures, as is the case in well managed newspaper publishing houses in Germany. The model does indeed require a critical discussion on journalistic and publishing optimization in terms of strategic perspective. In this context Meinhold discusses ten issues which should be considered in attempts to optimize the newspaper business.

The focus is on better exploiting the regional markets with a variety of initiatives which include new print concepts, online services and other offers. There are still hidden reserves in the business model with more professional sales, more efficient industrial organization as well as more economies of scope. Alternative financing models like non-profit organizations and cooperatives, on the other hand, are seen with skepticism. They pose dangers concerning the professional governance of publishing houses and the preservation of editorial independence.

6 The "shades of grey" in public relations

In the next segment, three institutional communication experts discuss recent (and not so recent) developments in PR currently affecting journalism. We'll begin with Klaus Kocks, former board member and corporate communications officer of Volkswagen, now a renowned PR consultant in Germany, followed by Barbara Baerns, professor emerita from the Freie Universität in Berlin, the doyenne of public relations research in Germany, and finally Marcello Foa, CEO of *Timedia* in Lugano, a veteran Italian journalist, foreign correspondent, and author of a remarkable book about spin doctors (Foa 2006).

As a practitioner, Kocks contributes remarkable insights. Trained in dialectics as well as in rhetoric, he describes how highly professionalized PR exercises significant public influence – without effects which could be measured easily with empirical research. Where others are engaging in "PR for PR" and in camouflage, Kocks addresses a provocative and refreshingly "honest" approach to his profession and mentions PR's "power of non-reporting," i.e. keeping the discussion about PR's influence out of the mainstream journalism agenda.

Barbara Baerns, who has focused her work on empirical analysis, emphasizes that the main objective for research in her field has always been "to expose latent relations and influences, and thereby create more transparent media coverage and a more transparent media system" – certainly a complementary perspective to Kocks'. Baerns declares the need to focus on online journalism and online PR to further analyze the changing relationship between the two professions. She wonders how her younger colleague Howard Nothhaft (University of Lipsia) could raise the question of "whether Barbara Baerns' determination thesis still deserves to be considered at all given the drastic changes in the media landscape" (Nothhaft 2012). Indeed, it is interesting to see what researchers – inspired by the desire to present some provocative thought – are willing to sacrifice during their rush to deflate another researcher's theory.

Marcello Foa bases his remarks on the distinction between "public institutional communication" which is supposed to be neutral and moderate, and "political communication" which tends to be partisan and biased. It is in itself interesting that such a distinction has survived in the Roman research culture which describes and analyzes, according to Hallin/Mancini (2005), "polarized-pluralist media systems" while the distinction is no longer made in the Anglo-Saxon world dealing with "liberal-market oriented media systems" or in Central and Northern Europe where a "democratic-corporatist media culture" prevails.

Foa's distinction between PR professionals who "act correctly applying licit techniques" (the good guys ...) and spin doctors who "aim not to inform but to manipulate media and public opinion" (the bad guys ...) is certainly a great starting point to analyze the "shades of grey" which exist between the two ideal types. His main argument is that journalists don't know enough about spin doctors to protect themselves and their publics from their subtle and sometimes not-so-subtle attacks. This certainly requires further reflection – not only among journalists themselves, but also among journalism educators who prepare future journalists for a media world where PR experts exert more influence than ever, and where journalism and PR converge (Russ-Mohl 2012).

Recent surveys and content analyses have sought to discover how journalists perceive the influence of PR and spin doctors (Weischenberg/Malik/Scholl 2006; Kerl 2007). They show that journalists are trapped by the tendency to underestimate the influence of PR and overestimate their abilities to deal with spin doctoring. Similar thinking has been discovered in Frenchmen: 84 percent of whom consider themselves masterly lovers. However basic laws of statistics dictate that there's only room for 50 percent to be above average, while by definition the other 50 percent can only end up below average with regard to love-making (Dobelli 2011: 14; concerning overconfidence see also: Vallone et al. 1990; Hoffrage 2004; Kahneman 2011).

In the case of journalists their overconfidence in dealing with PR adequately and to not become victims of spin doctoring is similarly understandable: The positive self-perception protects their egos. However, following Foa's short description of spin doctors' frequently used techniques, such overconfidence is naïve and frightening as well.

7 Complementary connections in the digital age

In the final section of the book, experts analyze the various ways search engines and social media affect journalism's future.

Christoph Neuberger, professor at the Ludwig Maximilians University in Munich, analyzes the relationship between journalism, social network sites, and news search engines. Referring to empirical findings, he states that "it is unlikely that journalism is in competition with social network sites and news search engines." Much more important are, according to him, the complementary connections between these fields – as social network sites and news search engines direct significant traffic to websites operated by print and broadcast media. Furthermore, newsroom staffs use social network sites and (news) search engines as research tools.

Thereafter Peter Laufer, a veteran American radio talk master and investigative journalist who recently became a journalism professor at the University of Oregon, and Oliver Quiring, a communications professor at and director of the Department of Communication at the University of Mainz, discuss crowdsourcing, swarm intelligence, and so-called "citizen journalism."

Laufer provides evidence that crowdsourcing has been practiced by his radio station during the San Francisco earthquake long before the advent of the Internet – and he strongly argues that "there is no need for journalists to rethink their reporting." Defending journalistic professionalism, he argues that "noise is not news." Random crowdsourcing of potential news stories should not be seen as a threat to traditional news reporting: "It is inane to assume such chatter should or would replace professional journalism." For Laufer, it remains unthinkable that the audience should dictate the contents of a newspaper or a broadcast. His credo: "It is the job of professional journalists to determine what is news," and: "Citizen journalism is an oxymoron unless that citizen happens to be a professional journalist."

Quiring concedes that according to research, few users actively contribute to the content pool on the Internet: "Users comment, share, and tag – but few are producing their own original news material." Nevertheless, he is convinced that journalists need to rethink their roles and should begin caring about user participation in Web 2.0 – as social media changes the process of news diffusion, as users are already dictating the content of media products, and as the endless flow and the wide diversity of raw news material on the Internet needs to be carefully curated by professional, independent journalists.

The question of whether we're disabled by experts (Illich et al. 2000) or whether, due to swarm intelligence (Surowiecki 2005) and empowerment by new technologies, experts are incapacitated by laymen may soon be yesterday's battlefield. Journalists and media companies must prepare for new challenges and the empowerment of the algorithm (Meckel 2011).

Ahmet Emre Acar's contribution at the conference moved in this direction. He represented the recently founded Humboldt Institute for Internet and Society in Berlin, funded by Google Inc. To underline how new technologies, in particular algorithms, affect journalistic work, he referenced the following examples:

- Thomson Reuters' *Calais Web Service*, a service that automatically attaches metadata to submitted content and links the submitted document with entities (people, places, organizations, etc.), facts, and events.
- *MemeTracker* which tracks the quotes and phrases that appear most frequently over time in online news and thus maps the daily news cycle so that everyone can see how successfully different stories compete for news and blog coverage day per day.
- *Digg* which pushes the most interesting news circulating in the Internet to the top and makes them "visible" for everybody.
- *Quora*, a question-and-answer-website, which connects people to everything they want to know by sharing content from the Web. *Quora* organizes people and their interests so that they "can find, collect and share the information most valuable" to them.

Acar referred extensively to a paper which Christopher W. Anderson (City University of New York) had prepared a few days earlier for the first Berlin Symposium on Internet and Society. In this paper, Anderson sheds light on how algorithms affect traditional journalistic work and ultimately, "the very definition of journalism itself" (Anderson 2011: 3). According to him, "journalists are only beginning to think about how algorithms might be used to manage their own informational workflows in a manner similar to that by which Google stores, retrieves, and ranks digital information on the entire Web." Anderson (2011: 5) mentions the need "to examine, in detail, the unequal distribution of computational resources in 21st century journalism" and to "critically dissect how this inequality is either impeding or facilitating journalism's professional mission."

As Anderson did not participate at our conference and Acar did not document his own contribution, our overview must stop here. From our point of view, the analysis suggested by Anderson would have to include the empowerment of new high tech monopolists like Apple, Google, and Facebook by unequally distributed computational and intellectual resources and by the data collections they own – which are, at least partially, as inaccessible as many government "secrets." However, straightforward mentions of this variety can probably not be expected from researchers generously sponsored by Google.

Stephan Russ-Mohl, professor of journalism and media management at the Università della Svizzera italiana in Lugano, Switzerland, and currently a Gutenberg Fellow of the Research Unit Media Convergence at the University of Mainz, concludes the book with final remarks – dealing with shortcomings of the media coverage of such diverse topics like the financial crisis, the Ehec virus, and the massacre in Norway, all of which have at least one fact in common: They demonstrate how poorly journalism is fulfilling its watchdog function.

Finally, the editors would like to express thanks to the Research Unit Media Convergence at the University of Mainz, in particular to its director and coordinator Stephan Füssel, and to Stefan Aufenanger, Dean of the Department of Social Sciences, Media and Sports, for having made the conference and this book possible, and to the partners of the European Journalism Observatory (EJO) in Lugano, Lucerne and Vienna, the MAZ – Die Schweizer Journalistenschule, and the Medienhaus Wien, personified by Marcello Foa and Natascha Fioretti, Sylvia Egli von Matt, Alexandra Stark and Daniela Kraus, for creative ideas they contributed during the planning phase of our conference and for their active cooperation in implementing them.

References

Anderson, Christopher W. Understanding the role played by algorithms and computational practices in the collection, evaluation, presentation, and dissemination of journalistic evidence. Conference draft. Paper prepared for the 1st Berlin Symposium on Internet and Society, 25th-27th October, 2011.

Ariely, Dan. *Predictably Irrational. The Hidden Forces That Shape Our Decisions*. New York: HarperCollins, 2008.

Asch, Solomon E. Effects of group pressure upon the modification and distortion of judgment. In *Groups, Leadership and Men*, Guetzkow, Harold S. (ed.), 177–190. Pittsburgh: Carnegie Press, 1951.

Cialdini, Robert B. *Influence: The Psychology of Persuasion*. New York: HarperCollins, 1998.

Dobelli, Rolf. *Die Kunst des klaren Denkens*. München: Hanser, 2011.

Erichsen, Jens. The new media ecosystem. Unpublished manuscript for the conference "Media Convergence & Journalism", University of Mainz, 21st-22nd October, 2011.

Foa, Marcello. *Gli stregoni della notizia*. Milano: Guerini e Associati, 2006.

Hallin, Daniel C./Mancini, Paolo. *Comparing Media Systems: Three Models of Media and Politics*. Cambridge: Cambridge University Press, 2005.

Hoffrage, Ulrich. Overconfidence. In *Cognitive Illusions: A Handbook on Fallacies and Biases in Thinking, Judgement and Memory*, Pohl, Rüdiger F. (ed.), 235–254. Hove: Psychology Press, 2004.

Illich, Ivan/Zola, Irving Kenneth/McKnight, John/Caplan, Jonathan/Shaiken, Harley. *Disabling Professions*. New York/London: Marion Boyars, 2000.

Kahneman, Daniel. *Thinking, Fast and Slow*. London: Penguin, 2011.

Kahneman, Daniel/Knetsch, Jack L./Thaler, Richard H. Fairness as a constraint on profit seeking: Entitlements in the market. In *Choices, Values, and Frames*, Kahneman, Daniel/ Tversky, Amos (eds.), 317–334. Cambridge: Cambridge University Press, 2000.

Kerl, Katharina. Das Bild der Public Relations in der Berichterstattung ausgewählter deutscher Printmedien. Eine quantitative Inhaltsanalyse. Unpublished Master's Thesis University of Munich, 2007.

Meckel, Miriam. *Next: Erinnerungen an eine Zukunft ohne uns*. Reinbek: Rowohlt, 2011.

Nothhaft, Howard. Rezension zu Ulrike Röttger, Joachim Preusse und Jana Schmitt: Grundlagen der Public Relations. Eine kommunikationswissenschaftliche Einführung. *Publizistik* 57, no. 2 (2012): 253–254.

Rainey, James. Los Angeles Times receives $1-million grant from Ford Foundation. Retrieved on 10th September 2012, from http://www.editorandpublisher.com/Newsletter/Article/Los-Angeles-Times-Receives--1-Million-Grant-From-Ford-Foundation (2012).

Ruch, Karl-Heinz. Personal talk with Stephan Russ-Mohl, 2nd May in Berlin, 2012.

Russ-Mohl, Stephan. *Kreative Zerstörung. Niedergang und Neuerfindung des Zeitungsjournalismus in den USA*. Konstanz: UVK Verlagsgesellschaft, 2009.

Russ-Mohl, Stephan. Opfer der Medienkonvergenz? Wissenschaftskommunikation und Wissenschaftsjournalismus im Internet-Zeitalter. In *Medienkonvergenz –Transdisziplinär*, Füssel, Stephan (ed.), 81–108. Berlin/Boston: De Gruyter, 2012.

Staun, Harald. Was genau war denn früher besser? In *Frankfurter Allgemeine Sonntagszeitung*, 29th July 2012.

Surowiecki, James. *The Wisdom of Crowds*. New York: Anchor Books, 2005.

Turner, Sebastian. How advertising changes in the digital revolution and how this affects news media. Unpublished manuscript for the conference "Media Convergence & Journalism", University of Mainz, 21st-22nd October, 2011.

Vallone, Robert P./Griffin, Dale W./Lin, Sabrina/Ross, Lee. Overconfident prediction of future actions and outcomes by self and others. *Journal of Personality and Social Psychology* 58, no. 4 (1990): 582–592.

Weischenberg, Siegfried/Malik, Maja/Scholl, Armin. *Die Souffleure der Mediengesellschaft. Report über die Journalisten in Deutschland*. Konstanz: UVK Verlagsgesellschaft, 2006.

Robert G. Picard
Killing journalism?

The economics of media convergence

Abstract: There will always remain a need for news and journalism. What is changing is the business of news: the ways that news have been financed, distributed, and consumed are being altered and the sustainability of existing news organizations has been altered; new technologies to produce and distribute news have appeared; and there is a shift in media use and a shift of power from the media to the consumers. News organizations must innovate and adapt to developments of the converging media world. Newspapers and news providers that cannot adjust to this new situation, that cannot effectively serve their original purposes, that can no longer serve the needs of their audiences, or that remain badly managed deserve to die.

Keywords: news organizations, financing, business models, change, innovation

---

1 Self-interest without self-reflection

The condition of journalism needs measured consideration, but that necessity is not being met through discussion among journalists today. Much of the dialogue is driven by newspaper journalists whose self-interests and beliefs that only existing newspaper-based organizations can provide quality journalism are clouding our understanding and limiting our consideration of other alternatives (Meyer 2004; Reilly Center 2008; Pickard/Stearns/Aaron 2009; Jones 2009).

News stories focus on declining revenue, layoffs, and bankruptcies in the newspaper industry and contain little of the business analysis traditionally given to other industries. There is little self-reflection and stories primarily illustrate problems appearing in a few locations, particularly the Anglo-American newspaper industries. The view conveyed is compounded by technophiles who are promoting digital media as an immediate substitution and go as far as saying that newspapers will be dead by 2015.

This has led proprietors of newspapers in Europe and North America to use the concern and rhetoric to their own advantage by trying to get beneficial subsidies, tax breaks, and protections of their business models.

There is no doubt that technological changes and the recent global recession have created enormous economic pressures on newspapers and other news organizations, but the underlying problem is that the traditional monopoly on news and information held by newspapers is gone. This change has been underway for several decades, but its effects have now become abundantly apparent.

2 Diminishing audiences, advertisers, and wealth

Today, radio, television and cable TV, Internet, mobile phones, and taxi, bus, and elevator screens all provide general news and information, business and finance news and analysis, sports news, analysis, and statistics, entertainment and celebrity news and gossip, and lifestyle news and features. The public is now using these platforms in ways that are altering their use of legacy news sources – especially newspapers and television news. The problem today is not that there is insufficient news and information, but that audiences are drowning in it.

All of these choices have led advertisers to alter their expenditure patterns (see Siegert in this book). Newspapers have paid a particularly heavy price – especially in countries such as the U.S. where papers had an unhealthy dependence on advertising income and relied on it for 85 percent of their income. The changes in advertiser choices and the downturn in the economy thus had dramatic consequences for the financing of newspapers. Even in Europe – where ad dependency tends to be 55 to 65 percent of income – the economic crisis produced significant declines in revenue (Currah 2009; Levy/Nielsen 2010).

These changes in the primary income stream of newspapers have been particularly painful because between 1950 and 2000 advertising revenues increased about 300 percent in real terms and made papers extraordinarily wealthy in the 1980s and 1990s, even as audiences were shifting their use to other media (Picard 2002). The last two decades of the twentieth century marked an unusual era for newspaper revenue, but today we think it was the norm for the industry throughout the last century.

As a consequence of the growing wealth, newspaper firms created large organizations with heavy overheads. They increased the size of the markets; they added journalists to their payrolls; they expanded their bureaus; they made great profits for the owners.

News providers of all kinds created expensive organizations and operations. News Corp. – which had revenues of 6.7 billion dollars (10.9 billion dollars today adjusted for inflation) in 1990 – reached revenues of 30.4 billion dollars in 2009. The BBC – which had an income of 220 million pounds (1.7 billion pounds today adjusted for inflation) in 1975 – had and income of 4.605 billion pounds in 2009 (27 times more money in real terms than in 1975).[1]

All the companies expanded and some commercial news firms, particularly in North America, took on heavy debt on the expectation the good times would continue (Picard 2006). They did not, of course, and the structures and costs developed in the wealthier era can no longer be maintained; hence, the cutbacks and downsizing.

A portion of the difficulties many firms are having adjusting to the changes in the contemporary media environment result from organizational success making change and innovation difficult. Growth of company size creates complexity and the need for more managers and greater division of responsibilities. This makes decision making more difficult and company policies and processes often inter- fere with the pursuit of new initiatives. This occurs because there is a natural tendency for companies to continue to follow paths that made them successful, especially if innovation often offers less short- to mid-term reward than historic activities.

This has been particularly troubling for the newspaper industry, whose iden- tity is integrally wrapped up in printing and traditional print journalism. The problem is that large portions of the public are now fleeing newspapers that once aggregated large audiences in some markets. Those who remain customers are spending less time with them, and young people are not widely adopting the newspaper reading habit.

This shifting media use is central to the challenges facing the news indus- try. The increasing competition from other types of media – in terms of number of media available and number of providers on those platforms – is wreaking havoc. Part of this trend is the result of changing society and lifestyles created by urbanization, commuting, changing roles of women, and new concepts of community and connectedness. In an era where technology empowers users, the desire and ability for increased control over content use and participation in the content itself are rising – a development that is anathema to many journalists (see Laufer and Quiring in this book). Audience members are no longer content to be spoon-fed news and information and are now exercising individual choices and developing new consumption habits (Picard 2011). Advertisers are following audiences and moving heavily to other forms of marketing – diminishing the role of advertising in their marketing mix.

---

1 Revenue figures are from News Corp. and BBC annual reports. Calculations of current dollar values were made by the author using U.S. Department of Commerce price inflator data.

Although these trends have particularly affected Anglo-American newspapers, if one looks behind the basic circulation and advertising expenditures in other nations a similar pattern is emerging and starting to have profound effects.

3 Shifting control, power, and organizational structures

Underlying all this is a power shift in communications. Media space was previously controlled by media; today it is increasingly controlled by consumers. News and information are no longer supply-driven, but demand-driven markets. The financing of all initiatives in cable and satellite TV and radio, audio and video downloading, digital television, and mobile media are based on a consumer payment model. Today, for every euro spent on media by advertisers, consumers now spend three euros. This has led companies to reduce advertising expenditures and they are now only about one-third of total marketing expenditures of major advertisers. Marketing money is moving to personal marketing, direct marketing, sponsorships, and cross promotions.

The shift in power is also facilitating production of consumer-created content with inexpensive and readily available creation software for audio, video, web design, personal sites, and blogging (Küng/Picard/Towse 2008). The technologies are promoting new types of news and information providers, peer-to-peer sharing, social networking, and sharing of information. Fixed place communications has shifted to mobile communications, and interactivity and user choices are shifting significant time to alternative interactive media uses (Bakardjieva 2005).

It may seem profane, but newspapers and news providers that cannot adjust to this new situation, that cannot effectively serve their original purposes, that can no longer serve the needs of their audiences, or that remain badly managed deserve to die. Journalists seem to think that newspapers have a right to exist and can exist forever. It is incredibly naïve to believe that any company or organization will last forever. Few companies last more than two or three generations. There are only about 1,500 firms worldwide that are two hundred years old. A few newspapers are on the list, but most are vineyards, breweries, hotels, and restaurants.

Economic, technical, and social trends are creating huge challenges for legacy news industries. The costs of news gathering and distribution are markedly lower for digital media and these lower costs of entry facilitate risk-taking in developing new means of conveying news and information, and allowing the introduction of novel news and information products by other firms. The new products are revealing product limitations and flaws in legacy news media, and weaknesses in customer orientation of legacy media, thus attacking economics of scale in legacy media, and making organizational inefficiencies and cost structures even more untenable. Change is no longer a choice; it is a requirement for news organizations!

Some important evidences of changes are: the increasing disputes among owners and board members of legacy media; newspapers selling their large city-centre buildings; reorganizations and break ups of newspaper groups; the closure and divestiture of less profitable or central activities; unsolicited offers to purchase news organization assets; and new owners and funders entering the industry with different motives and operational ideas.

But does this mean that journalism is being killed? Listening to the rhetoric one would certainly think it is.

If one steps back, however, it is clear that there will always remain a need for news and journalism. The need for news is not changing; what is changing is the business of news. The ways that news has been financed, distributed, and consumed are being altered and the sustainability of existing news organizations has been altered. The tempo of change is affecting news media differently. Some countries are being affected more rapidly than others. Broadcast audiences are more affected than print audiences, but the financial situation of newspapers is most affected because they have the highest non-news cost structures. National and large metropolitan newspapers are most affected in some countries, but local newspapers are most affected in other countries (Picard 2010).

The appearance of new means of communication has always induced changes in existing media by altering their functions and roles, their business models, and their use, forcing them to adapt to survive. In the past two centuries, telegraph services changed with the spread of the telephone, radio altered when television appeared, television channels adjusted to development of cable, and newspapers are now changing because of television, cable, and the Internet.

But change does not necessarily lead to wholesale destruction. We still send the short quick messages of the telegraph, albeit by SMS and e-mail. We listen to the radio for entertainment, news, sports, and discussion of public affairs and watch television for the same purposes. We read newspapers for news, analysis, and features. We use the Internet for communication, social interaction, information, and news. One is not completely replacing the other but is changing the amount we use the various media and creating new preferences for serving certain types of content from particular media.

But we should not really be concerned with the fate of particular media; what should concern us is the fate of journalism. We thus need to think carefully about the future and how we respond to the changing conditions. The big question is: "How do we finance and organize journalism in the twenty-first century?"

Obviously, news organizations will need smaller and more agile operations and take a far more entrepreneurial approach than in the past. New and different types of news and supplemental news providers will emerge to fill gaps. News organizations of all types will have to become more innovative in their products and processes and will have to cooperate through alliances and networking in ways they have been reticent to do in the past. Revenue will have to come from multiple sources and reduce the dependency on any one source. It the end, managers will have to rethink their entire business models for media to ensure that it creates greater value for customers.

4 Journalism is separate from news enterprises

Questions about whether we are witnessing the end of journalism have made many journalists in legacy news operations apprehensive. But much of the concern results from a misunderstanding of the nature of journalism.

Journalism is not a job; it is not a company; it is not an industry; it is not a business model; it is not a form of media; it is not a distribution platform. Journalism is an activity. It is a body of practices by which information and knowledge is gathered, processed, and conveyed. The practices are influenced by the form of media and distribution platform, of course, as well as by financial arrangements that support the journalism. But one should not equate the two.

The pessimistic view of the future of journalism that many hold is based in a conceptualization of journalism as static, with enduring processes, unchanging practices, and permanent firms and distribution mechanisms. But if one considers history, one sees that journalism has constantly evolved to fit the parameters and constraints of audiences, media, companies, and distribution platforms.

In its first centuries, journalism was practiced by printers, part-time writers, political figures, and educated persons who acted as correspondents – not by professional journalists as we know them today. In the nineteenth century, the pyramid form of journalism story construction developed so that stories could be cut to meet telegraph limits, and production personnel could easily cut the length of stories after reporters and editors left their newspaper buildings. Professionalism in the early twentieth century emerged with the regularization of journalistic employment, and professional journalistic best practices developed. The appearance of radio news brought with it new processes and practices, including "rip and read" from the news agencies teletypes and personal commentary. TV news brought a heavy reliance on short, visual news, and 24-hour cable channels created practices emphasizing flow-of-events news and heavy repetition.

Journalistic processes and practices have thus never remained fixed, but journalism has endured by changing to meet the requirements of the particular forms in which it has been conveyed and by adjusting to resources provided by the business arrangements surrounding them. Journalism may not be what it was a decade ago – or in some earlier supposedly golden age – but that does not mean its demise is near. Companies and media may disappear or be replaced by others, but journalism itself will adapt and continue.

It will adjust not because it is wedded to a particular medium or because it provides employment and profits, but because its functions are significant for individuals and society. The question facing us today is not whether journalism is at its end, but what manifestation it will take next. The challenges facing us are to find mechanisms to finance journalistic activity and to support effective platforms and distribution mechanisms through which information can be conveyed.

5 The need to look forward

The perception that convergence and its economic effects are killing journalism can be attributed to an unfortunate and very human tendency to think in simplistic terms. There is a tendency to think of the glorious past and that the future can never be as good. There is a tendency to think there can be only one winner in a competitive market and that gain for one must come at the expense of another. These ways of thought have obscured understanding of what is happening to journalism and news media and what it means to society.

The changes are most felt in newspapers today because the mass media business model that nurtured them for nearly a century is becoming less viable. Television news is also being devastated. Reliance on advertisers to pay the vast majority of the costs is no longer viable because the mass audience is changing into a niche audience that is less attractive to many advertisers.

I do not wish to minimize the shock, dislocation, and harm that the changes have caused for journalists and other employees at newspaper firms and other news organizations. They are very real and very unpleasant. The kind of decline we have witnessed in the past four years is particularly devastating to firms with high overheads, unreasonable debt loads, and investors clamoring for high dividends – an apt description of many news firms before the downward plunge.

The challenges confronting news organizations are not trivial or to be diminished, but we need to view them knowledgeably to realistically assess their real effects, and to avoid panic. Journalism can survive the storm that is pounding news organizations if we maintain the will for it to survive, if we take decisive action to promote journalism in new forms, and if we do not endanger it by poor choices. The process will not be comfortable, but the storm will subside, and innovative news organizations will appear beside legacy news firms. The journalists who work in them will hopefully emerge wiser and seek to avoid mistakes that increased their vulnerability to this storm.

News organizations will probably never return to the halcyon days that made them so attractive to commercial investors in the last decades of the twentieth century. However, if they are run effectively by reasonable proprietors with sensible overheads and debts, they will be able to carry on and to serve the journalistic needs of society for many years.

References

Bakardjieva, Maria. *Internet Society: The Internet in Everyday Life*. London: Sage, 2005.

Currah, Andrew. *What's Happening to Our News: An Investigation into the Likely Impact of the Digital Revolution on the Economics of News Publishing in the UK*. Oxford: University of Oxford, Reuters Institute for the Study of Journalism, 2009.

Jones, Alex S. *Losing the News: The Future of News that Feeds Democracy*. Oxford: Oxford University Press, 2009.

Levy, David A.L./Nielsen, Rasmus Kleis. *The Changing Business of Journalism and its Implications for Democracy*. Oxford: Reuters Institute for the Study of Journalism, University of Oxford, 2010.

Küng, Lucy/Picard, Robert G./Towse, Ruth. *The Internet and the Mass Media*. London: Sage, 2008.

Meyer, Philip. *The Vanishing Newspaper: Saving Journalism in the Information Age*. Columbia, MO: University of Missouri Press, 2004.

Picard, Robert G. *Evolution of Revenue Streams and the Business Model of Newspapers: The U.S. Industry between 1950–2000*, Discussion Papers C1/2002, Business Research and Development Centre, Turku School of Economics and Business Administration, 2002.

Picard, Robert G. Capital crisis in the profitable newspaper industry. *Nieman Reports* 60, no. 4 (2006): 10–12.

Picard, Robert G. The future of the news industry. In *Media and Society*, Curran, James (ed.), 365–379. London: Bloomsbury Academic, 2010.

Picard, Robert G. *The Economics and Financing of Media Companies*. 2nd ed. New York: Fordham University Press, 2011.

Pickard, Victor/Stearns, Josh/Aaron, Craig. *Saving the News: Toward a National Journalism Strategy*. Washington, D.C.: Free Press, 2009.

Reilly Center for Media and Public Affairs. *The Breaux Symposium: New Models for News*. Baton Rouge: Louisiana State University, 2008.

Gabriele Siegert

From "the end of advertising as we know it" to "beyond content"?[1]

Changes in advertising and the impact on journalistic media

Abstract: The advertising industry and the media industry have long been tied together to reach their main objectives. The advertising industry used media as ad vehicles to embed and transport their ad messages and the media needed advertising money to finance and subsidize their activities. Additionally the advertising income of media outlets depends on economic changes – be they cyclical or structural. Journalistic media seem to be more affected by cyclical downturns than other media types, and they seem to be at least as much affected by structural changes than other media. Structural changes in advertising as well as the possibility to combine advertising in new ways, lead to a loss of advertising money for journalistic media. While advertising money is still important in the financing of journalistic media, at the moment the future of this funding source is unclear. Most likely, advertising revenue will not be large enough to finance newsrooms that are designed to make important contributions to democratic societies.

Keywords: interactive advertising, integrated advertising, personalized advertising, conversation and recommendation as ad objectives

---

1 Introduction

The advertising industry and the media industry were long bound together to reach their main objectives. On the one hand, the traditional revenue model of the media was highly dependent on advertising income. Media outlets needed advertising money to finance and subsidize their journalistic activities. On the other hand, the advertising industry needed the media to serve as a vehicle to embed and transport their advertising messages – in order to achieve the traditional objectives of advertising, such as gaining attention or interest.

---

[1] The title refers to two studies of the IBM Institute for Business Value (Berman/Battino/Shipnuck/Neus 2007 and Berman/Battino/Feldman 2010).

Of course both industries have always tried to meet their targets by using new and different measures. The advertising industry has tried to be successful by using direct marketing, below-the-line advertising, or program integrated advertising to reach their target groups. The media industry has tried to finance their products partly via fees paid by the audience or via income from other sources. Nevertheless to achieve their main objectives – attention and consumption in order to attract big audiences and big money – the advertising industry and the media industry were tied together. This mutual interdependence was addressed by media economists as *two-sided-markets* (e.g. Rochet/Tirole 2006). Is this mutual interdependence changing due to media convergence?

Starting from the assumption that the explained mutual interdependence is still the basis of media production, I will define the main underlying concepts, "journalistic media" and "media convergence," in two short paragraphs. Thereafter I will explain the changes in advertising mainly along the dimensions objectives, messages, formats, and vehicles. I also refer to some issues, which Sebastian Turner and Jens Erichsen, both leading professionals in the German advertising agency industry, raised during the conference on which this book is based (see Russ-Mohl/Nienstedt/Wilczek in this book). Finally, I will discuss the impact of these changes on the media industry and try to answer the question "Where is advertising heading – and what is the future for journalistic media as an advertising vehicle, while media convergence progresses?"

2 Journalistic media

When discussing the impact of changes in advertising on the media, it is important to realize that we are not talking about the media or media performance in general. As the book title points out clearly, the problem is about journalistic media. The ongoing public discussion on the future of the media as a result of the financial crisis in 2008 and 2009 is also not dealing with media performance in general but with journalistic media. The funding of journalism is the problem that academia and parts of the public are concerned about. We are worried about journalistic media, more precisely, media outlets who make important contributions to the functioning of democratic societies. These media outlets serve an important role in their ability to convey accurate political information, attract public attention, and to a certain extent – exercise control over political, economic, and business issues and actors. Hence it is the journalistic media that is most important for society, and it is the journalistic media outlets who are most susceptible to changes in the way that advertising funds journalistic content.

The advertising income of the media in general is dependent on economic changes to the market – be it cyclical or structural changes. Past research has clearly shown that the advertising income of media outlets is dependent on cyclical fluctuations (e.g. Grimes/Rae/O'Donovan 2000; Picard 2001; Chang/Chan-Olmsted 2005; Lamey/Deleersnyder/Dekimpe/Steenkamp 2007; van der Wurff/Bakker/Picard 2008). There is also empirical evidence that not all media types react similarly to cyclical downturns. Media outlets that are active in journalism, particularly newspapers, seem to be more affected than others. Their advertising income reacts much stronger to cyclical downturns of the GDP, then for example, electronic media (Picard 2001 for the U.S. and Europe; Shaver/Shaver 2005 for Asia; van der Wurff/Bakker/Picard 2008; Deleersnyder/Dekimpe/Steenkamp/Leeflang 2009). Hence the journalistic media suffer more from recessions than other media.

A recently finished project on the structural change of advertising (Siegert/Kienzler/Lischka/Mellmann 2012) finds evidence that consumer spending on durable consumer goods (DCG) as well as advertising expenditures of DCG providers are more closely related to the business cycle than consumer spending on selected fast moving consumer goods (FMCG) and advertising expenditures of FMCG providers, respectively. In this respect, media companies should have an eye on their advertising customer because private consumption of DCG might serve as a feasible predictor for future advertising income. Additionally there is also evidence that print media is affected more by business cycles than other media (Lischka/Kienzler/Siegert 2012).

Additionally the advertising income of the media is dependent on structural changes in the advertising industry. Our previously mentioned project shows that the structural change of advertising is not a short-term, rapid change caused by economic downturns and recessions. Rather it has to be seen as a long-term, gradual process pushed but not caused by economic downturns and recessions. Again not all media types are affected in the same way – though it is not easy to prove empirically. There are however, arguments to support the thesis that journalistic media is affected at least as much as other media.

3 Media convergence

Media convergence is said to be the overall change agent – for the media as well as for the advertising industry. Key-drivers of that change are interactivity (cues are: user-generated content, "produsers," that is users of online texts, pictures, videos or music turning also into producers and distributors of texts, pictures, videos or music, networks, transactions, communities, affiliate-marketing, long-tail), integration (cues are: infomercials, advertorials, placements, advertiser founded – not just funded – programming in terms of branded entertainment) and personalization (cues are: one-to-one-marketing, mass customization, behavioral targeting) (Siegert/Brecheis 2005; Berman/Battino/Shipnuck/Neus 2007; IAB-PWC 2009; Siegert 2010).

Although the basis of the media and advertising change is technological, what really makes a difference for the industries is the change in media and ICT usage. Unfortunately precise numbers of ICT users vary due to the applied source (Eurostat, Internet World Stats, comScore), but it is without doubt that the usage of the Internet and mobile services is rising rapidly worldwide. Amongst the 27 EU-member countries, the percentage of Internet users raised from 36 in the year 2004 to 65 in the year 2010.[2] Following Internet World Stats (March 2011) the penetration rate of the Internet (in percentage of the population) worldwide looks as follows: Africa 11.4 percent, Asia 23.8 percent, Europe 58.3 percent, Middle East 31.7 percent, North America 78.3 percent, Latin America/Caribbean 36.2 percent, Oceania/Australia 60.1 percent.[3]

In addition, Internet users increasingly use online content for information and orientation purposes. Online offers are therefore a new competitor for journalistic print media in the audience market. "Large percentages of Internet users in most of the WIP (World Internet Project, GS) countries and regions go online to seek local, national, or international news. In all of the reporting countries and regions other than Colombia, more than 25 percent of users go online to look for news at least daily, and more than half go online for news at least weekly" (World Internet Project 2010). Unfortunately this increasing usage is not in line with increasing revenues for content and news producers. This seems to be due to changes in the functioning of advertising, as well as changes to the way advertising is using online and mobile platforms.

---

2 Statistics on the information society by Eurostat: http://epp.eurostat.ec.europa.eu/portal/page/portal/information_society/data/main_tables (retrieved on 26th July 2012).
3 Internet World Stats (March 2011): www.internetworldstats.com. Copyright © 2001–2011 Miniwatts Marketing Group.

4 Changes in advertising

The way advertising works changes due to media convergence and modified media usage. In this regard the different dimensions of advertising change can be structured as follows:

– *Changes in advertising objectives:* Advertising was and still is persuasive communication particularly in its attempts to influence knowledge, attitudes and behavior. However, conversation and recommendations from third parties are new or at least additional advertising objectives in convergent times. Older models of mass media effects, such as the two-step-flow of communication (Lazarsfeld/Berelson/Gaudet 1948; Katz/Lazarsfeld 1955) take conversation and conversation networks into account. In recent years the topic became attractive again for communication science researchers (e.g. Southwell/Yzer 2007 and 2009; Yzer/Southwell 2008). Communication experts see democratization in brand management due to the possibilities of online communication ("produsers"). Advertisers hope for conversation about their advertisements without having knowledge on "who talks when where and why positively or negatively about advertising" (Wiencierz 2012). To involve the brand in conversations and to achieve recommendations, advertisers do not necessarily need journalistic services, because conversations and recommendations (word of mouth) are based on interpersonal and network communication. For interpersonal and network communication they need technologies, platforms or ecosystems, not journalistic media in the traditional sense of the meaning.
– *Changes in advertising messages:* Editorial thinking becomes crucial for the communication of brands, because irresistible ideas and content are drivers for any kind of conversation. Producing content, editorial thinking, and storytelling are core competences of traditional media – for both journalistic and entertainment media. In particular, media outlets that specialize in entertainment produce their content in close co-operation with the advertising industry. They develop magazines, stories, shows and series that can easily be combined with advertising messages from the corresponding industry. This type of collaborative communication enables companies to act as centers of competence concerning particular topics, be it media companies or companies from other industries. Therefore the connection between media and brand communication seems to be non-detachable. Unfortunately brands start to create their own content and thanks to new technologies they have ways to distribute it to desired target groups. What started decades ago

with corporate publishing is now a content creation of its own – viral videos are an example as well as magazines such as the *Red Bull Bulletin*.

- *Changes in advertising formats:* Looking at the history of advertising, we find immense creativity in inventing new advertising forms and formats – an ongoing differentiation. This differentiation of advertising can only be roughly outlined with direct marketing, below-the-line advertising (ambient and ambush marketing), or program integrated advertising. PR and advertising are merging, and so are the formerly separate formats: "Just what is integrated marketing communications? ... it appears to be the natural evolution of traditional mass-media advertising, which has been changed, adjusted, and refined as a result of new technology" (Schultz 1999: 337). Additionally, advertising and entertainment are merging and are leading to new hybrid advertising formats such as product placement, advertorials or in game advertising. The new formats again are not necessarily connected to journalistic media – a lot of them can stand alone. Format and measures such as viral marketing, word-of-mouth, mobile advertising, promotion, sponsoring, direct marketing, and customization are used increasingly (Grimes/Rae/O'Donovan 2000; Perez-Latre 2007; Arora et al. 2008).
- *Changes in advertising vehicles:* In general, advertisers try to find the best mix of advertising vehicles to meet their objectives. Due to media convergence and the upcoming possibilities to reach the desired target groups, companies start to change their advertising investment behavior. The Internet experienced a considerable increase in advertising income in the past (e.g. Linnett 2002; Saksena/Hollifield 2002; Siegert 2010), in particular regarding classifieds (Picard 2008). Traditional media as paid advertising vehicles are beginning to be seen as old fashioned. The new forms and formats promise to reach a mass audience or specific target group without the high costs of media buying. The moment retailers begin looking for other advertising channels – which is currently taking place – traditional media will lose a substantial amount of advertising revenue. Even though retailers will continue to use traditional media in order to distribute promotion material directly to households, the practice is not necessarily connected to the production of journalistic content or the quality of that content.

The advertising-supported Internet is becoming the most important competitor of journalistic media, specifically regarding advertising money. This is not only due to the immense amount of advertising money directed towards the Internet already, it is also due to the fact that the Internet offers new possibilities that function without media content and therefore pose serious challenges to journalistic media as advertising vehicles.

The advertising-supported Internet refers broadly to all activity on the Web intended to promote marketplace exchange of products, services, or information. Paid online advertising is one component. In addition, most e-commerce websites perform a substantial information and promotional function, to encourage commerce. Therefore, e-commerce providers can be thought of as Internet advertisers. Many websites that do not conduct e-commerce also perform an informational advertising function. On behalf of both for-profit and not-for-profit enterprises, they take the place of magazine advertising, brochures, and direct marketing, educating the consumer on features and benefits of the organization's offerings. Additionally, e-commerce sites and company websites collect data about customers and prospective customers. They perform an interactive advertising function analogous to sales forces. So, too, do e-mail solicitations, another form of interactive advertising on the Internet. Internet-enabled economic activity is a dynamic system. New advertising methods, such as the development of paid search in 2003, have expanded the amount advertising contributes to funding the Internet. (Hamilton Consultants/Deighton/Quelch 2009: 3 f.)

The communication process of advertising as a whole is modified due to key-drivers of media convergence – interactivity, integration and personalization. One-way communication is out of style. A brand's task is to create programs of activity that allow customers to engage whenever they want – 365 days per year. Two-way communication and interactivity are not core competences of traditional journalistic media. Referring to Maletzke's definition, mass communication is an indirect, one-way communication (Maletzke 1963: 32), which usually does not allow for participation and interaction. Although it is obvious that currently many companies are still unable to cope with the openness and the two-way communication of the Internet or critical online word of mouth, they will adjust to it sooner or later, and they will change their advertising investment behavior at the disadvantage of journalistic media.

Furthermore, rising mobile media usage is addressed by technologies, not by journalistic content. Location based news is predominately not the type of news defined in the beginning of this article. Without having empirical data on this issue it's apparent that the average mobile media user is not particularly interested in journalistic content. For example a customer who is exposed to news about a war, while walking through the city and going shopping. Location based news is focused on a different kind of content – especially, when we think of the combination of advertising, content, commerce, social networks, and mobile communication. An example of the typical combination could be an online diet program with special recipes (content) that are automatically combined with a list of ingredients linked to a range of products (advertising) from a particular retailer, with an option for the customer to buy or order (commerce). The list could be either sent to the computer or to the mobile phone of the recipient (mobile communication). A diet blog and recipe section would serve as the area for recommendation and word of mouth – social networks would additionally allow for improving the service. This combination of advertising, content, commerce, social networking, and mobility focuses on people's real life and their everyday problems. Usually this kind of content does not match with the traditional issues journalism deals with.

In summary, the changing way advertising works and uses online and mobile technologies is affecting the traditional media altogether – most of all the journalistic media.

5 Impact on the media industry – an outlook

I will try to answer the question about the future for journalistic media as an advertising vehicle by referring to Ian R. Bruce (1999: 473). He points out that advertising decisions are most often founded on the question "Will this sell?", whereas public relations decisions are often founded on the question "Will this help establish beneficial relationships?"

However the questions advertising decisions are founded on will change in the following way:

1) *Advertising is still designed "to sell":* When the traditional role of media as a vehicle for advertising is addressed, it is due to the distribution capacity of traditional media outlets, and to their reach of a mass audience or particular target groups. Some brands will no longer need journalistic media to reach their desired target groups, but there will still be many products – from low fat yogurt to beer – which need to reach an audience that is not continuously online. Although people increase their online and mobile usage, media consumption as a whole is not changing so dramatically. In many countries, TV is still the advertising vehicle with the highest reach. Therefore it still makes sense to advertise low fat yogurt in the context of "Desperate Housewives" and beer during the "Champions League" final.

2) *Advertising is no longer used only "to sell," but also "to entertain":* Brand communication needs new ideas, original views of the world, and interesting stories. It wants not only to inform and persuade it also wants to entertain. When it comes to storytelling and entertaining, traditional media still plays an important role. Brands need the media to embed or to integrate their story and their ideas in an advertising-friendly content. In this regard, entertainment media is more important for brand communication than journalistic media. But all traditional media will suffer from the fact that increasingly

brands create their own stories, their own content, and are able to distribute it to desired target groups by using new channels.

3) *Advertising is no longer used only "to sell," but also "to talk about":* New advertising objectives such as conversation and recommendation make the advertising industry prefer new technologies and platforms beyond journalistic content. However, in many countries traditional media monopolizes the process in order to promote public conversation. Online activity might initiate the process – but without the engagement of journalistic media it will stay online and not reach the general public. If advertisers want a broader target group (including not only the young urban people who are continuously online) or the public as a whole to talk about a brand or an issue, they need traditional media, including journalistic media – at least to pick up the idea.

4) *Advertising is no longer used only "to sell," but also "to interact and participate":* Traditional media as producers of journalistic content are not meant to be direct and interactive, neither concerning the underlying and preferred technology nor concerning the production process or the producers, the journalists. Especially when it comes to direct connection of content with commerce, online and mobile technologies are the better and more efficient choice. Online and mobile technologies might be combined with content, but what content would that be? There is more than one reason that this would be entertaining content or "news you can use" instead of in-depth "stressful" news about politics or the economy.

What is the future for journalistic media as an advertising vehicle then? In the past, the differentiation of advertising vehicles and advertising possibilities has already reduced advertising income for traditional media. The upcoming changes in advertising due to media convergence will certainly continue this process for all traditional media. In 2007, the IBM Institute of Business Value saw the role of traditional media in the advertising value chain already shrinking. Although they concentrate on broadcasters and the United States, their arguments are convincing:

> Arguably, broadcasters that rely on linear television advertising to fund operational and content costs are at risk in a world of increasing consumer control, niche content and fragmented attention. And yet, broadcasters have the opportunity to leverage their current mindshare with customers, while transforming their operations to embrace the plethora of new digital content distribution opportunities. By delivering integrated, crossplatform advertising programs tied to their programming assets, they can migrate into a successful future model. (Berman/Battino/Shipnuck/Neus 2007: 16)

Therefore the IBM Institute of Business Value named their study with reference to the foreseeable structural change "The end of advertising as we know it." As journalistic media is affected by cyclical downturns more than other media types, recessions could accelerate this structural change. Altogether, the structural change of the media industry is closely connected to the structural change of the advertising industry (see Figure 1).

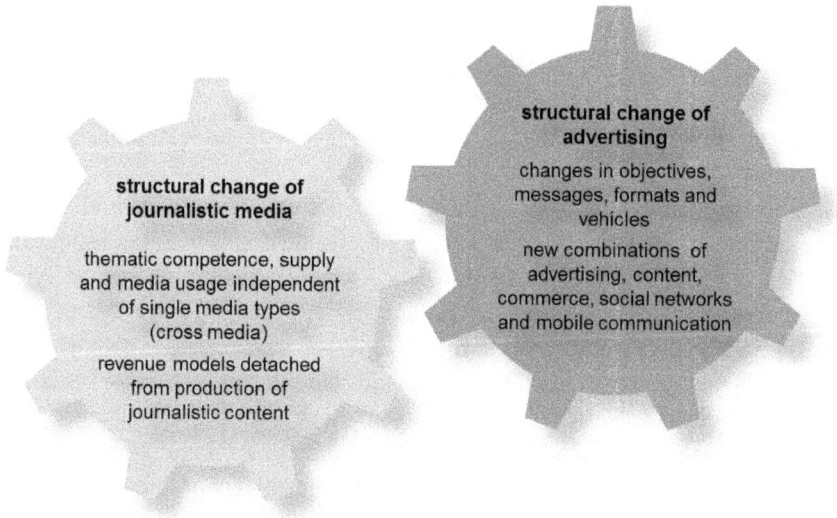

structural change of advertising

changes in objectives, messages, formats and vehicles

new combinations of advertising, content, commerce, social networks and mobile communication

structural change of journalistic media

thematic competence, supply and media usage independent of single media types (cross media)

revenue models detached from production of journalistic content

Figure 1: Structural change of advertising and journalistic media. (Source: own Figure).

Meaning that in the future, there will be some advertising investments in journalistic media the question is whether this spending, this piece of advertising money, will be big enough to finance the type of journalistic media that informs, controls as watchdogs, and acts as a forum for political debates. Serious doubts are shown because advertising-supported journalistic media is facing extensive changes due to media convergence. As a result of the mutual interdependence between advertising and the media industry, "the end of advertising as we know it," means "the end of the revenue model of journalistic media as we know it."

Hopefully journalistic media will never develop according to the title of the 2010 IBM study "Beyond content" (Berman/Battino/Feldman 2010), because democratic media and information societies depend on the "journalism" as an institution. Democracy cannot get away with the decline of journalistic media – just because advertisers might find better distributors for their messages.

References

Arora, Neeraj/Dreze, Xavier/Ghose, Anindya/Hess, James D./Iyengar, Raghuram/Jing,
    Bing/Joshi, Yogesh/Kumar, V./Lurie, Nicholas/Neslin, Scott/Sajeesh, S./Su, Meng/Syam,
    Niladri/Thomas, Jacquelyn/Zhang, Z. John. Putting one-to-one-marketing to work:
    personalization, customization, and choice. *Marketing Letters* 19, no. 3/4 (2008):
    305–321.
Berman, Saul J./Battino, Bill/Shipnuck, Louisa/Neus, Andreas. The end of advertising as we
    know it. IBM Institute for Business Value. Retrieved on 21st November 2007, from http://
    www-935.ibm.com/services/us/gbs/thoughtleadership/?cntxt=a1000062 (2007).
Berman, Saul J./Battino, Bill/Feldman, Karen. Beyond content. Capitalizing on the new revenue
    opportunities. IBM Institute for Business Value. Retrieved on 22nd August 2010, from
    http://public.dhe.ibm.com/common/ssi/ecm/en/gbe03361usen/GBE03361USEN.PDF
    (2010).
Bruce, Ian R. Public relations and advertising. In *The Advertising Business. Operations,
    Creativity, Media Planning, Integrated Communications*, Jones, John Philip (ed.), 473–483.
    London/New Delhi: Thousand Oaks, 1999.
Chang, Byeng-Hee/Chan-Olmsted, Sylvia M. Relative constancy of advertising spending. A
    crossnational examination of advertising expenditures and their determinants. *Gazette,
    The International Journal for Communication Studies* 67, no. 4 (2005): 339–357.
Deleersnyder, Barbara/Dekimpe, Marnik G./Steenkamp, Jan-Benedict E.M./Leeflang, Peter S.H.
    The role of national culture in advertising's sensitivity to business cycles: An investigation
    across continents. *Journal of Marketing Research* 46, no. 5 (2009): 623–636.
Grimes, Arthur/Rae, David/O'Donovan, Brendan. Determinants of advertising expenditures:
    Aggregate and cross-media evidence. *International Journal of Advertising* 19, no. 3 (2000):
    317–334.
Hamilton Consultants/Deighton, John/Quelch, John. Economic value of the advertising-
    supported Internet ecosystem. Retrieved on 10th September 2012, from http://www.iab.
    net/media/file/Economic-Value-Report.pdf (2009).
IAB-PWC. IAB Internet advertising revenue report conducted by PricewaterhouseCoopers (PwC).
    Retrieved on 17th September 2009, from http://www.iab.net/insights_research/530422/
    adrevenuereport (2009).
Katz, Elihu/Lazarsfeld, Paul F. *Personal Influence*. Glencoe, Il: Free Press, 1955.
Lamey, Lien/Deleersnyder, Barbara/Dekimpe, Marnik G./Steenkamp, Jan-Benedict E.M. How
    business cycles contribute to private-label success: Evidence from the U.S. and Europe.
    Journal of Marketing 71, no. 1 (2007): 1–15.
Lazarsfeld, Paul F./Berelson, Bernard/Gaudet, Hazel. *The People's Choice. How the Voter Makes
    Up His Mind in a Presidential Campaign*. New York: Columbia University Press, 1948.
Linnett, Richard. Magazines pay the price of TV recovery. *Advertising Age* 73, no. 35 (2002):
    1–2.
Lischka, Juliane/Kienzler, Stephanie/Siegert, Gabriele. Advertising expenditures, consumer
    spending and business expectations. Assessing these relations for German data from
    1991–2009. Presentation accepted for the European Communication Research and
    Education Association (ECREA) Conference 2012, "Social media and global voices",
    24th-27th October 2012, Istanbul, TR, 2012.

Maletzke, Gerhard. *Psychologie der Massenkommunikation*. Hamburg: Hans Bredow-Institut, 1963.

Perez-Latre, Francisco J. The paradigm shift in advertising and its meaning for advertising-supported media. *Journal of Media Business Studies* 4, no. 18 (2007): 41–49.

Picard, Robert G. Effects of recessions on advertising expenditures: An exploratory study of economic downturns in nine developed nations. *Journal of Media Economics* 14, no. 1 (2001): 1–14.

Picard, Robert G. Shifts in newspaper advertising expenditures and their implications for the future of newspapers. *Journalism Studies* 9, no. 5 (2008): 704–716.

Rochet, Jean-Charles/Tirole, Jean. Two-sided markets: A progress report. *RAND Journal of Economics* 37, no. 3 (2006): 645–667.

Saksena, Shashank/Hollifield, C. Ann. U.S. newspapers and the development of online editions. *International Journal on Media Management* 4, no. 2 (2002): 75–84.

Schultz, Don E. Integrated marketing communications and how it relates to traditional media advertising. In *The Advertising Business. Operations, Creativity, Media Planning, Integrated Communications*, Jones, John Philip (ed.), 325–338. London/New Delhi: Thousand Oaks, 1999.

Shaver, Mary A./Shaver, Dan. Changes in the levels of advertising expenditures during recessionary periods: A study of advertising performance in eight countries. Paper presented at the Asian-American Academy of Advertising, Hong Kong, 2005.

Siegert, Gabriele. Online-Kommunikation und Werbung. In *Handbuch Online-Kommunikation*, Schweiger, Wolfgang/Beck, Klaus (eds.), 434–460. Wiesbaden: VS Verlag für Sozialwissenschaften, 2010.

Siegert, Gabriele/Brecheis, Dieter. *Werbung in der Medien- und Informationsgesellschaft. Eine kommunikationswissenschaftliche Einführung*. Wiesbaden: VS Verlag für Sozialwissenschaften, 2005.

Siegert, Gabriele/Kienzler, Stephanie/Lischka, Juliane/Mellmann, Ulrike. Medien im Sog des Werbewandels. Konjunkturell und strukturell bedingte Veränderungen der Werbeinvestitionen und Werbeformate und ihre Folgen für die Medien. Report. (Project funded by the Swiss National Foundation), 2012.

Southwell, Brian G./Yzer, Marco C. The roles of interpersonal communication in mass media campaigns. In *Communication Yearbook 31*, Beck, Christina (ed.), 420–462. New York: Routledge, 2007.

Southwell, Brian G./Yzer, Marco C. (eds.). Conversation and Campaigns. Communication Theory 19, no. 1 (2009): 1–101.

van der Wurff, Richard/Bakker, Piet/Picard, Robert G. Economic growth and advertising expenditures in different media in different countries. *Journal of Media Economics* 21, no. 3 (2008): 28–52.

Wiencierz, Christian. An expert's view on conversations and advertising. Presentation at the Advertising in Communication and Media Research Symposium 14th-15th June 2012, University of Tuebingen, 2012.

World Internet Project. World Internet Project report finds large percentages of non-users, and significant gender disparities in going online. Retrieved on 10th September 2012, from http://www.digitalcenter.org/WIP2010/wip2010_long_press_release_v2.pdf (2010).

Yzer, Marco C./Southwell, Brian G. New communication technologies, old questions. *American Behavioral Scientist* 52, no. 1 (2008): 8–20.

Part 2: **Financing Journalism in the Future**

Heinz-Werner Nienstedt, Bettina Lis

More money from media consumers

Paid content and the German newspaper case

Abstract: German newspapers have more time to adapt to convergence and the digital age when compared to the United States, because their core print business is in a healthier position than U.S. newspapers. Nevertheless, to be able to finance quality journalism in the long run German publishers have to create new business models for their digital publications. As the poor performance of advertising revenues may prove to be a long-term trap, paid content models are the focus of discussions in this respect. This paper discusses reasons why paid content models make sense and which challenges exist when attempting to introduce them.

Keywords: newspapers, financing, business models, paid online content, pricing

---

1 Performance and structural differences of newspapers in the U.S. and Germany

When one looks to the reader's side of the market, newspapers are a growing medium on a global level. According to Wan-Ifra (2010), the daily circulation of paid newspapers grew by roughly 6 percent between 2005 and 2009, however sharp differences exist between markets. While newspaper circulation in Africa (+30 percent), Asia (+20 percent) and South America (+5 percent) grew, it shrunk by 11 percent in North America and by 8 percent in Europe. Meanwhile newspapers lost significant market share in advertising in the same amount of time. On a global level, newspaper share of overall advertising expenses went down during the same time span from 27 percent to 21 percent (PwC 2008, 2010) with the market share winner being the Internet.

Emerging newspaper markets operating elsewhere are of little help for U.S. or European publishers. Newspapers cannot be exported to growing regions with minor exceptions, and even concerning investments in these regions opportunities are limited because of a number of market entry barriers. This is a major difference to industries like cars, chemicals and others. Newspaper publishers therefore have to fix their problems in their home markets.

As the U.S. market has been the focus of recent news concerning the ongoing newspaper crisis, we will first compare the essential components of the market to the German situation in order to better understand the different challenges the German newspaper industry faces. We'll mainly look at numbers from 2005 – before the financial crises began and the industry had recovered to some extent from the recession of early 2000 – to 2009 where numbers are available on a comparable basis.

U.S. markets turned out to be the eye of the hurricane during the recent newspaper crises. As can be seen in Table 1, print newspaper advertising lost 48 percent (or 22.6 billion dollars) from 2005 to 2009 (with a loss of 7.3 billion dollars in classified ads) and reached only 24.8 billion dollars in 2009. The market share of newspaper advertising in all of U.S. advertising expenses decreased to 14 percent and in the next two years added a decrease of another 8 percent and 9 percent respectively. The online ad revenues could not balance this loss. It grew only by 0.7 billion between 2005 and 2009. Circulation of newspapers shrank at a much smaller pace than advertising, 15 percent during the same time period. Price increases helped that circulation revenues were even less affected. They fell by 6 percent (all numbers from NAA 2012).

Table 1: Comparison of key figures in newspaper industries in the U.S. and Germany. (Source: *NAA 2012; **BDZV 2007, 2010).

		2005	2009
U.S.*			
	Print advertising	47.408 billion dollars	24.821 billion dollars
	Online advertising	2.027 billion dollars	2.743 billion dollars
	Circulation volume	53.345 million	45.653 million
	Circulation revenues	10.747 billion dollars	10.067 billion dollars
Germany*			
	Print advertising	4.729 billion euros	3.903 billion euros
	Circulation volume	27.40 million	25.31 million
	Circulation revenues	4.215 billion euros	4.473 billion euros

U.S. publishers cut their editorial staff by more than 22 percent in the same time period, resulting in an newsroom staff of about 40,000 today (cf. Pew Research Center's Project for Excellence in Journalism 2012). In the 1980s and the 1990s there were about 55,000, 38 percent more than today.

In Germany newspapers got less into trouble compared to the U.S. Advertising fell by 17 percent between 2005 and 2009 to 4 billion euros resulting in a loss of market share of 2.7 percent. Nevertheless newspapers remained the largest advertising medium with 22 percent, followed by TV (20 percent market share in 2009). This is the case although the level of advertising sales is now roughly back on the level of the end of the 1980s when East Germany was still not included. Circulation decreased by 8.3 percent, but circulation income grew by 6 percent due to copy price increases (see Table 1).

Editorial staff (excluding trainees ("Volontäre") whose number increased by about 200 in the same time) was only cut by 4 percent or 660 people to about 14,000. The number of newsroom staff nevertheless equaled roughly the 1996 figure where newspapers had been in very good shape. That means that newspaper publishers did not significantly reduce editorial staff during the last years development, since another 5 percent reduction of editorial staff happened in 2010 (all numbers for Germany: BDZV 2007, 2010).

It is interesting to compare the numbers of newsroom staff in the U.S. with the numbers in Germany, in relation to their population. The relation of journalists between Germany and the U.S. was 1:2.7 in 2009 and had been 1:3.8 in 1993 according to the above-mentioned sources. The relation of the population served is 1:3.9 today. This highlights the steep decline in journalistic resources in the U.S. due to the most recent layoffs, which may have affected the value provided to readers in contrast to Germany.

There are further structural differences between the U.S. and Germany. Newspaper reach in 2009 was much higher in Germany with 71 percent of adults older than 14 years of age – compared to the U.S. where the number was only 44 percent (NAA 2012). We find evidence of declining audience reach dependent on the age of readers in Germany, but the low average from the U.S. is only met by very young cohorts between 14 and 19, being 42 percent. Even 53 percent of those between 20 and 29 years old read or scroll through newspapers on an average daily basis and those between 40 and 49 meet the average German reach of 71 percent (BDZV 2010). Long-term research has shown stable reading behavior from cohorts over their lifetime. If that would be true over future decades, the average reach would decrease to only 67 percent according to a simulation of ZMG (2011). Unfortunately there are traps when conducting such simulations due to rather sudden fluctuations of digital media consumption habits, making markets difficult to forecast. In addition, newspapers having older generations as the dominant readers' segment may lose their general attractiveness to younger readers much faster – and even more to advertisers. To gain those publics as customers, who prefer online only news consumption and to maintain those who integrate digital media in their media portfolio, adequate digital offers have to be made.

The German newspaper industry is not as dependent on advertising as U.S. newspapers. In the U.S. the portion of revenues from circulation was 18 percent in 2005, and then numbers increased to 29 percent in 2009 due to the steep decline of advertising (cf. Pew Research Center's Project for Excellence in Journalism 2012). Respective numbers for Germany were 46 percent and 53 percent. The difference is mainly due to much higher purchase prices, which start at 0.25 dollar in the U.S. and 1 euro in Germany for non-tabloids (Wan-Ifra 2010). Traditionally, U.S. newspapers rely strongly on the audience making function of the reader's side of the two-sided market, subsidizing this side and receiving their financing mainly from the advertiser. This too is the case in Germany's digital world today, with mostly free online offers and pure ad financing – something which should be seen as a warning sign if one is willing to learn from the development of U.S. print markets.

2 Strategies for profit growth in the digital age

In summary, German newspapers – specifically regional newspapers, which represent 90 percent of the circulation of non-tabloid dailies – hold a much healthier market position than the U.S. newspaper industry. In addition, some newspaper publishing houses have diversified their activities to find new growth potential and sources of revenue, mostly by making use of their own core competences and resources, e.g. content generation, knowledge about local customers, or logistics. Examples are free advertising-driven weeklies delivered to all households, targeted ad-driven magazines, phone directories and postal services. Developments in this field will continue and all of them can help to maintain corporations' profitability even if revenues from the core of the print newspaper businesses dwindle.

In both countries, purchase price increases served as a major source of revenue growth from consumers. These prices eased as most newspaper sales in both countries are paid through subscriptions. In Germany, the effects were stronger than in the U.S. due to a higher portion of revenue from circulation as well as higher copy prices, however price elasticity for newspaper subscriptions is low (cf. Wirtz 2001: 116). Although further incremental real price increases may be possible without severely increasing churn, some segments of readers with a lower willingness to pay will be lost if prices are raised significantly. In the case of news magazines in Germany, price elasticity of potential subscribers was found to be much higher than the elasticity of existing subscribers (Gieseking 2010). If that also holds true for newspapers, price increases would make it more difficult and more expensive to find a new generation of subscribers, thus substituting those who drop out, meaning there are limitations to that strategy of revenue and profit growth.

To maintain the core business of news in the mid and long term, publishers face the challenge of adapting to changing media usage patterns and to the new rules of the advertising market in the digital age. German publishers may have more time to adapt to these challenges in comparison to the U.S. due to their healthier situation. But this time is necessary, since no reasonable business model has been found for the digital news market until now.

3 Rethinking the current online news business model

Over the last years the Internet has evolved into a more mature phase. In Germany, total reach was 73 percent in 2011 with more than 95 percent between the ages of 14 and 29 years old. Average usage time per day is 137 minutes, a usage time that has remained at this level since 2009 (ARD-ZDF 2011). This may also serve as a reason to rethink the Internet strategy and its business model.

Nearly all newspapers have cultivated branded websites and an increasing number serve mobile smartphones and tablets. Their revenue model for news on the Web is predominantly advertising only. While published figures cannot be found on the success or failure of this model in terms of revenues and profits, publishers' hopes for a prosperous future using this model vanished over the last years, which can be heard in numerous statements from conferences and on other occasions. With the possible exception of some national market leaders or of publishers who diversified their digital business in other fields like e-commerce, we find most news providers and especially regional newspapers in a situation where revenues at best cover the marginal costs of online activities or make a positive margin on a low level of revenues resulting in small profits. Marginal costs mean that only incremental costs of online news outlets are allocated to this business. Costs of the print editorial team, where most of the news basis for the website is produced as well as other organizational costs from the publishing house, are in general not allocated to the online activities.

The reasons for the growing skepticism of the pure advertising model are twofold. First, it is a more realistic view on the size of the relevant online advertising market. The share of total online advertising spent in Germany in 2011 has been estimated to be 19.6 percent by OVK (2012), the organization of online advertising marketers, representing 3.3 billion euros in revenues. Online overtook newspapers, which represented only 18.6 percent of the market according to this source. But these are gross numbers, they in addition also include search, which is mainly Google, and affiliate advertising. Looking only at classical online advertising (banners, pre-rolls, etc.) and excluding these two items, which are relevant to news sites only to a very low extent, and also taking net numbers, which include discounts, ZAW (2012), the association of the advertising industry, estimates online advertising to be 0.99 billion euros with a market share of 5.2 percent in 2011. Daily newspaper share in contrast is at 18.8 percent. A large portion of this 0.99 billion euros is absorbed by the vast amount of sites other than news driven ones, and especially by sites of big players and aggregators with traffic much higher than those of news media and/or are disseminated to websites independently from their content by targeting techniques. This leaves a more limited volume available to news publishers than expected, in case one starts from the huge gross numbers.

The large difference between gross and net numbers roughly represents discounts are about 70 percent for classical online ads according to these statistics, indicating that the market conditions are an advantage for the buyer. It can be expected that the strength of the buy side will further increase with an ever growing inventory of ad space on the Internet resulting in decreasing CPMs. Therefore expectations of strong midterm growth from advertising revenues are also limited.

This may have caused the German publisher Hubert Burda (2009) to speak about "lousy pennies" only, which can be achieved from the online advertising market by publishers. A pessimistic, maybe realistic view is that the growth in online advertising will cover the increase of further marginal costs derived from the publishers' online efforts. But it will not be able to cover the basic editorial resources which feed online but are allocated to print today. In the long term this would be necessary when news consumption further migrates to digital media.

4 Paid content models as a potential solution

On the look out for alternatives to poor ad financing, paid content models for online news are the focus of many projects and discussions from publishers today. For a long time it was seen as impossible to bill users for news on the Internet, where content is overwhelmingly free. The *Wall Street Journal* and the *Financial Times*, who have run successful paid models for years, were long seen as outstanding exceptions. Their ability to bill users could be reasoned by the "need to know" content of financial news, something that is not applicable to general newspapers.

Especially *The New York Times'* success of gaining roughly 500,000 digital subscriptions in the year after the introduction of the metered model in 2011, where more addicted users who surpass a threshold number of articles in one month can do so only when they become subscribers, created more optimism for the potential success of paid content models of general newspapers, prompting hundreds of U.S. newspapers to experiment along this line (cf. VDZ 2012). In Germany some publishers like *Axel Springer AG* and *Madsack* have shown to be forerunners, with outlets utilizing both subscription and pay per view models for premium content, especially in local news.

In addition to finding new revenue sources there are other reasons which may persuade publishers to turn to the paid content model on the Web. For starters, there is an intrinsic trap of free news websites whether they provide limited or comprehensive content in comparison to print. Many newspaper sites do not present their best and exclusive content, including background stories or the full volume of local content or commentaries, on the Web – they leave those to the paid print edition of their brand. If Web content is limited in such a way, there will be a dilution of the brand and its value for those who only use the online arm of the newspaper brand – with negative long term effects on the ability to introduce premium pricing or even pricing at any level. If news coverage is on the other hand as comprehensive as print is and in addition offers more actuality and functionality, there is a clear danger of cannibalization of the paid print edition. Why should all those who are experienced enough to use the Web pay a premium price for the newspaper in this case? They may deem the newspaper to be more comfortable to read, but to get the same news for free is a highly attractive choice. In addition, valuable paying newspaper readers have the right to feel bad if they pay and others get the same for free, which may diminish their loyalty.

Last but not least there is an attractive further value proposition for paid content, one that turns IP-addresses into customers, thus fostering opportunities for learning about their preferences. In addition, increased interaction increases the chance for CRM practices based upon the rich information that Internet usage provides within the legal framework, probably leading to additional sales and higher customer value. Under the umbrella of a customer contract data protection contract, rules are different and interaction is eased. In addition, better known customers and customers who prove to be loyal and express their valuation towards the online content and the brand by their subscription may be more valuable for many advertisers.

5 Reasons for resistance

In spite of the advantages of paid content models, resistance to implement them is still strong in many publishing houses. Several reasons and hurdles for this can be addressed. We will summarize and discuss them here.

At first there are reasons which have their roots within the organization of publishing houses. Online knowledge is often concentrated within online departments, meaning editors and upper management rely on their experience and judgment. These online departments have operated to maximize traffic and to increase advertising revenues for years now. After a long period of doubt about whether digital formats make sense for newspapers, except as a marketing tool for print and after tentative phases of expansion and detraction in online departments, these eventually have achieved a relative stability due to results which are just above break even or still slightly negative but acceptable to management. Paid content strategies raise the fear that traffic, and as a consequence online advertising revenues, is negatively affected and losses of ad revenues will not be balanced by additional income from consumers. This may lead to new phases of instability and threats to the appreciation of online departments, their resources, and activities – a fear that possibly causes their opposition to the introduction of paid content models. Indeed it is a risk to change a business model and change must be guided by general strategies for which top management is responsible. It is their duty to decide in favor of a change if they believe in paid content strategies. In this case, support from their online specialists in the information and decision phase may be limited, which is a difficult situation for the decision-making process. Many U.S. publishers took the decision in favor of paid content and subsequent promising results from recent experiences should be a hint, that the risk of losing traffic and advertising revenue is limited to none existent if paid content models are chosen with care (VDZ 2012).

As a major source of risk, the question of whether consumers are at all willing to spend money for news can been identified. This question has different aspects: First there has always been a normative string of argumentation, that consumers should oppose paying, since the nature of the Internet is free content. Recently, in the context of piracy and intellectual property rights, these arguments were offended more heavily by the claim of artists, especially those engaged in music, film and books as well as publishers and distributors claiming to have the right to earn a fair amount for their work. At least this discussion led to broad public contemplation of these issues. Online activists championing for free content share a sort of general understanding of these arguments when they are put forward by artist, but much less so for equivalent arguments from media companies. In the case of newspapers, where author rights are predominantly waived to the publisher, the "enemies" are the media companies which are perceived by these groups to be dispensable in the digital age anyway. Therefore they have a challenging task to achieve in overcoming the "right for free information" mentality.

A second argument is that it is impossible to implement successful paid content models because one always finds for free substitutes on the Web. This statement is attached to the question of exclusivity of content. It systematically underestimates the amount of exclusivity that most newspaper content provides. For example, the profoundness of news content, the extent of local news, the analysis and explanation as well as the commentary, and also the individual tonality and political angle in which the news is presented by editorial teams. In addition we find trust and specific associations towards newspaper brands, which go beyond the content value. All this leads to the high level of print newspaper purchases that are made each day, although the Internet and its superabundance of news and information is there.

A third argument is that news have the character of public goods as soon as they are published (cf. Picot 2009; Kaye/Quinn 2010: 12). This is a question of executing property rights on the Internet, provided sufficient efforts are made to exclude at least a relevant majority of non payers by technical measures. This is clearly much more difficult than in the analogue world but not as difficult for news outlets as for blockbuster films, music, or novels, where consumer interest lasts longer and audiences are much bigger. In these cases it is much more attractive to take the risk of a systematic breach of intellectual property rights.

Another argument is that users should not be asked to pay because of the structure of the market (cf. Dewenter 2009: 657). In the context of a two-sided market structure, it may be better to take the financing from the advertising market. Neglecting marginal costs, the rational is that one side of the market should take the burden of financing a platform such as a news site, which has a lower price elasticity and lower positive externalities than the other side of the market (cf. Rochet/Tirole 2003). The latter is clearly true for the advertising side in the case of news sites, where the users' side has higher positive externalities than the advertising side. According to this, news media may therefore function as purely non-paying audience makers for advertisers. The question is whether the price elasticity of the advertising side, in relation to the consumer side, balances this effect. No empirical studies to answer that question are currently known, specifically in the context of online markets. Low CPMs and increasing discounts are an indication that the price elasticity of the advertising market is quite high. Anyway the results of the ad financing model have not proven to be successful in real life.

The question that remains is whether consumers would be willing to pay for news on the Internet if asked to do so. Previous polls about the willingness of consumers to pay have resulted in a wide range of data which generally depends on the sample and the questions asked – mostly a direct question ("would you be willing ... "). The results of such polls therefore vary a lot, e.g. different studies for Germany suggest values for those who are willing to pay for news on the Internet between low 6 percent (GfK-Verein 2009) and high 67 percent (BCG 2009). These polls therefore hardly provide reliable guidance when determining what publishers can expect from consumers.

6 Pricing for news content on the Internet

Indeed it is difficult to introduce a price for what has previously been consumed for free for years. Consumers have anchors in their mind when contemplating a price point for a product they are willing to purchase. These anchors may come from various sources and may be accidental (cf. Ariely 2009: 27 ff.). In the case of news on the Internet the anchor price can be assumed to be zero because users have been trained by publishers to think that this is the case. Zero is more than a normal anchor, it is an emotional hot button, a category of its own where consumers tend to forget the possible downside of a transaction and where consumers choose products which are not in their best interest over products where the price may be low (cf. Ariely 2009: 55 ff.). This makes it hard to bill for news services when free alternatives exist but at a lower quality. Users must be convinced of the distinctiveness and exclusivity of the paid product, which is a challenge for publishers when communicating the quality of the product.

Moving upwards from zero is a price increase. Research in behavioral economics provides hints about circumstances which support consumer acceptance of price increases. The positive message is that consumers not only look at their own self-interest but also at the legitimate interests of the supplier. According to the Dual-Entitlement-Principle developed by Kahneman/Knetsch/Thaler (1986a, 1986b) consumers start their judgments from a reference point and account for the disadvantages of buyers and the motives of suppliers. Supplier demands are eventually accepted, but only if they are deemed to be fair. Research about why purchasers accept a price increase detected drivers and prohibitions for fairness. Prohibitions found include the motive to increase profits or to wield power as a result of a monopoly. "Good motives" in contrary lead to the perception of price fairness, which includes compensation for rising costs or for preserving jobs (cf. Kahneman/Knetsch/Thaler 1986a: 734 f., 1986b: 295 ff.; Campbell 1999). Since fairness is a central point when inducing the willingness to pay, publishers have to convince users that the introduction of paid content meets their perception of fairness.

To date, existing research is limited to helping publishers in this respect since there has been no research about fairness drivers in the context of media, including news sites. In addition, previous research has dealt primarily with price increases but not from a starting price point of zero. Therefore we recently performed an empirical study in order to fill this gap (Kopp/Nienstedt 2012).

Throughout the course of an adaptive conjoint analysis, roughly 500 test persons voted on the introduction of paid regional newspaper sites which differed concerning 25 items belonging to a set of seven categories (time of introduction, allocation of the revenues, business rational, communication policy, discounting policies, tariff system, amount of advertising). For the study, such categories and items were chosen that were expected to influence the perception of fairness. Items with the lowest utility values, i.e. prohibitions of fairness, have been lack of communication of reasons, lack of a specific goal for the allocation of revenues, advertising similar to advertising found on free sites, increase of profits, introduction of paid content as first mover. The biggest drivers of perceived fairness turned out to be preserving jobs/preventing a shrinking news staff, investment into more creative options for young journalists, proper and early communication, test subscriptions, and freeness from advertising. In the next step, where willingness to pay was tested by the price sensitivity meter method, it was found that expected maximum revenues could be roughly doubled when each individual received a set of items which was chosen to be fair in his view over an offer deemed to be unfair. This indicates the high importance of fairness aspects when introducing paid content models.

A crucial question is the price to be demanded for paid content news sites. On the basis of a further conjoint experiment (Nienstedt/Ebel 2012) with 800 test persons, all users of news websites – drawn from subscribers and online users of a national newspaper brand as well as from a convenience online sample – we could simulate the buying decisions of the test persons at different price points for monthly subscriptions. In the conjoint they were offered their preferred online news brand as well as three others chosen randomly from a list of prominent online news sites in Germany. They also could choose the non-option, meaning they would not choose any of the offered but rather another free-content news site. The products offered varied concerning the brand, price, portion of content offered for free, access (Web or iPad app), and specific additional content and functionality. The analysis of relative importance of these constructs showed that price was most important followed by the brand while all other constructs had only minor importance. Looking at utilities, one could see that the app had a lower utility than the Web.

In a simulation on the basis of the estimated preferences of the conjoint analysis, we forecasted the buying decisions of the users. In cases where the app of the preferred brand was at 9.99 euros, and the website of the preferred online brand as well as two other randomly chosen prominent online brands varied in their monthly subscription price, about half of the users turned to the non-option at a monthly price of 2.99 euros for the three websites. At a price of 9.99 euros, slightly more than 70 percent switch to the non-option and would probably use other free news sites. The preferred brand garners a 50.7 percent share of preference for its app and its website when its website is free and its app costs 9.99 euros. The share drops by about 60 percent to 20.6 percent when the website is also at 9.99 euros. In terms of revenues, in case the preference share is converted to real buying decisions, this means that if a free newspaper website has 100,000 users for whom it is the preferred brand and then loses 60 percent of them, it would still generate about 4 million euros in revenue for its Web and app. The heavy loss of users would be compensated by significant revenues, probably more than it could earn through advertising, meaning the result would encourage the introduction of a paid subscription model at a fairly high price, in this case 9.99 euros. This result shows a strong market potential – however, it will take time and efficient marketing efforts to unlock this potential. This experiment shows the probable effects of a "hard pay wall" for a limited amount of Web content.

The recent praxis and introduction of the metered model in the U.S. indicated that after a downturn of traffic in the first months, the total traffic recovered again to its old level in short time according to industry sources cited by VDZ (2012). Total revenues from subscribers of print and digital editions are said to have increased by 4 percent to about 6 percent while advertising income was not negatively affected according to these sources. Given the short time in which these results were achieved, the introduction of the metered model appears promising.

Practical experiences in the U.S. as well as our conjoint results for the German market show that paid content models are more a chance than a risk for publishers. Even if revenues for online sites are not increased significantly in the short term, such paid content models help to change the price anchor and the reference price for news content. This allows the preservation of the traditional mixed revenue model of newspapers in upcoming times where media consumption will progressively move to digital. It may also help to recruit younger and online-only audiences as paying consumers in the future. Findings from behavioral economics research show that it will take time to change the price anchor of consumers. To start with a low price policy could be a trap in this respect, since low prices indicate low quality and can hardly be changed to premium prices. Whatever the right pricing scheme is – be it metered models or paid models for selected content (so called freemium models) – it has to be chosen specifically for each publication.

In addition, the decision is normally not about one singular price point, but rather about specific menus which offer print, e-paper, website, smartphone app, and tablet applications as choices for consumers. This means that product and price bundling strategies have to be applied. Optimizing the price structure for individual products as well as bundles is a new task for publishers. By apply-ing proven methods to this problem from other industries, such as fast food chains and the automotive industry, a great deal of insight can be applied when approaching these issues. Thus pricing to consumers turns out to be a field in which publishers need a new and more sophisticated level of professionalism.

We also showed that to be successful it is essential that consumers accept the paid model to be fair. Further efforts in research and praxis should be allocated to elaborate ways in order to achieve that. To turn consumers' free lunch of online news into a successful paid model which can help to finance quality journalism is turning out to be a long term task. German newspaper publishers can hope-fully be happy to have more time to accomplish that task than U.S. publishers, as explained above. Their time should not be wasted.

References

ARD-ZDF Onlinestudie. Retrieved on 10th September 2012, from http://www.ard-zdf-onlinestudie.de/ (2011).

Ariely, Dan. *Predictably Irrational. The Hidden Forces That Shape Our Decisions*. New York: HarperCollins, 2009.

Boston Consulting Group (BCG). Press release – news for sale: charges for online news are set to become the norm as most consumers say they are willing to pay, according to The Boston Consulting Group. Retrieved on 10th September 2012, from http://www.bcg.com/media/PressReleaseDetails.aspx?id=tcm:12-35297 (2009).

Bundesverband Deutscher Zeitungsverleger e.V. (BDZV). *Jahrbuch Zeitungen 2007*. Berlin: ZV Zeitungs-Verlag Service GmbH, 2007.

Bundesverband Deutscher Zeitungsverleger e.V. (BDZV). *Jahrbuch Zeitungen 2010*. Berlin: ZV Zeitungs-Verlag Service GmbH, 2010.

Burda, Hubert. Conference: DLD (Digital – Life – Design). Munich, 2009.

Campbell, Margaret C. Perceptions of price unfairness: antecedents and consequences. *Journal of Marketing Research* 36, no. 2 (1999): 187–199.

Dewenter, Ralf. Die Preissetzung von Internet Content Providern: Oder naht das Ende der „Gratiskultur"? Zeitgespräch: Was darf das Internet kosten? *Wirtschaftsdienst* 89, no. 10 (2009): 657–659.

GfK-Verein. Die Bereitschaft, für Internetinhalte zu bezahlen, ist gering – Internationale GfK-Studie zur Internetnutzung in 17 Ländern. Retrieved on 13th September 2012, from http://www.gfk.com/imperia/md/content/presse/091211_wsje_internet_dfin.pdf (2009).

Gieseking, Thomas. *Gewinnoptimale Preisbestimmung in werbefinanzierten Märkten. Eine conjoint-analytische Untersuchung eines Publikumszeitschriftenmarktes*. Wiesbaden: Gabler, 2010.

Kahneman, Daniel/Knetsch, Jack L./Thaler, Richard H. Fairness as a constraint on profit seeking: Entitlements in the market. *The American Economic Review* 76, no. 4 (1986a): 728–741.

Kahneman, Daniel/Knetsch, Jack L./Thaler, Richard H. Fairness and the assumptions of economics. *Journal of Business* 59, no. 4 (1986b): 285–300.

Kaye, Jeff/Quinn, Stephen. *Funding Journalism in the Digital Age. Business Models, Strategies, Issues and Trends.* New York: Peter Lang, 2010.

Kopp, Sven/Nienstedt, Heinz-Werner. Wie wichtig ist wahrgenommene Preisfairness für die Akzeptanz der Einführung kostenpflichtiger journalistischer Nachrichtenangebote im Internet? Working Paper Medienmanagement Mainz, 2–2012.

Newspaper Association of America (NAA). Trends and numbers. Retrieved on 13th September 2012, from http://www.naa.org/Trends-and-Numbers.aspx (2012).

Nienstedt, Heinz-Werner/Ebel, Fridtjof. Pricing für App und Online: Eine conjointanalytische Untersuchung überregionaler Nachrichtenangebote. Working Paper Medienmanagement Mainz, 1–2012.

Onlinevermarkterkreis im Bundesverband Digitale Wirtschaft e.V. (OVK). Online-Werbeinvestitionen nähern sich 2011 der 6-Milliarden-Euro-Grenze. Retrieved on 13th September 2012, from http://www.ovk.de/ovk/ovk-de/online-werbung/datenfakten/ werbeinvestitionen-nach-segmenten.html (2012).

Pew Research Center's Project for Excellence in Journalism. The state of the news media 2012 – an annual report on American news media. Retrieved on 13th September 2012, from http:// stateofthemedia.org/ (2012).

Picot, Arnold. Erlöspolitik für Informationsangebote im Internet. Zeitgespräch: Was darf das Internet kosten? *Wirtschaftsdienst* 89, no. 10 (2009): 643–647.

PricewaterhouseCoopers (PwC). *Global entertainment and media outlook: 2008–2012, Industry Overview, 9th annual edition.* New York, 2008.

PricewaterhouseCoopers (PwC). *Global entertainment and media outlook: 2010–2014, Industry Overview, 11th annual edition.* New York, 2010.

Rochet, Jean-Charles/Tirole, Jean. Platform competition in two-sided markets. *Journal of the European Economic Association* 1, no. 4 (2003): 990–1029.

Verband Deutscher Zeitschriftenverleger (VDZ). New Media Trends & Insights. Berlin, 2012.

Wan-Ifra. *World Press Trends 2010 Edition.* Darmstadt, 2010.

Wirtz, Bernd. *Medien- und Internetmanagement,* 2nd ed. Wiesbaden: Gabler, 2001.

Zeitungs Marketing Gesellschaft mbH & Co. KG (ZMG). Standortbestimmung Junge Leser: Desktop-Research – Eine erste Grundlagenstudie zum Thema im Auftrag des BDZV. Frankfurt am Main, 2011.

Zentralverband der Deutschen Werbewirtschaft (ZAW). Medien: Die meisten im Plus. Retrieved on 13th September 2012, from http://www.zaw.de/index.php?menuid=119 (2012).

Konny Gellenbeck
Investing in "return on political income"

How the *tageszeitung* was saved from bankruptcy

Abstract: The German newspaper *tageszeitung* (*taz*) survived financial crises because it managed to turn its readers into associates and its customers into publishers. From the very first issue in the late 1970s, the *taz* was the result of solidarity among its readers. And since this very beginning, its readers felt responsible for the financial survival of their news source. So when the *taz* turned into a non-profit cooperative in the early 1990s to avoid bankruptcy, several thousand joined. Meanwhile, the *taz* continues to tap into various digital communication channels by sending newsletters, posting on Facebook, and tweeting headlines, in order to offer the most personal experience for its readers and thereby fostering valuable relationships.

Keywords: journalism, business models, cooperative, relationship management, social network sites

1 With a little help from its readers

The German newspaper *tageszeitung* (*taz*) is owned by its readers, a business model which has nurtured the success of the publication. The paper was born as an alternative national daily newspaper in 1978 by a group of young activists from various movements, including the women's rights, anti-nuclear, and peace movements.

Most of the founders were not journalists but they agreed that traditional newspapers were failing to appropriately cover what they saw as important current events. From a publisher's point of view, the *taz* should not have been able to survive a single year, however in a political sense such a project was overdue. The founders unveiled their idea in a brochure and offered customers the chance to subscribe to a newspaper that did not yet exist, a newspaper that would be different from those already available. Before long the founders of the *taz* realized that they were not alone in their search for a newspaper with topical variety and a unique self-organization. 7,000 readers eventually embraced the opportunity to subscribe and paid for an annual subscription.

Therefore from the very first issue, the *taz* was the result of solidarity among its readers. The myth about its foundation was the cornerstone for all levels of cooperation to come.

Customers who initiated their subscription agreements, blindly entered into a very concrete and personal relationship with those producing the newspaper. From the very beginning, the readers of the *taz* felt responsible for the financial survival of their newspaper, and in the early stages it was extremely necessary. Due to chronic underfunding, each small financial setback presented an enormous threat to the existence of the newspaper – despite having a staff which was strongly motivated by intrinsic goals and already "exploiting" itself by working for low salaries. In the first ten years of its existence, the *taz* started several "rescue campaigns" which asked customers to not only buy and read the newspaper but also to subscribe in order to assure its liquidity.

2 The cooperative arises

Fourteen years later in 1992, a cooperative emerged from the self-governing business and in 2012 more than 12,000 people ensure the economic and journalistic independence of their newspaper. After the fall of the Berlin Wall, the German newspaper-market changed dramatically. The struggle for keeping the *taz* alive seemed lost. In 1991, the newspaper faced either the threat of bankruptcy or the possibility of being sold to a large publishing company. From this financial crisis emerged a plan to convert the newspaper into a cooperative, by turning readers into associates and *taz* customers into publishers.

Within four months, 3,000 readers joined the cooperative by paying a one-time minimum deposit of 500 euros, but many paid more. At the conclusion of their founding-year, the *taz*-cooperative accumulated 2.5 million euros in cash reserves, and today their capital stock is worth 10 million euros and growing.

Part of the allure of *taz* is its non-profit status, making reader and associate support for their newspaper uniquely enduring. One associate explained her commitment to the newspaper by referring to its contents as "political income." Associates are well aware that the offer the *taz* is providing is not designed to adhere to the market, but towards journalistic relevance. Those who praise the paper's unbendable and independent attitude are more likely to help the publication financially.

Along the way readers learned the basic business rules of the publishing sector:

- Subscriptions are like small credits, but free from interest.
- Scope is as important as sales figures.
- Traditional newspapers received 80 percent of their financing through advertisements.
- The price of paper varies.
- Wholesale traders do not stock every kiosk with the *taz*, and so forth.

3 A community based on solidarity

The communicative foundation and repeated calls for solidarity were designed to explain these publishing sector rules, while additionally providing readers with the opportunity to choose between three different pricing options for their subscription. Those who had more to spend were encouraged to subsidize a subscription for other customers. Currently there are many readers paying the higher "political price" for the *taz*, thus securing the subsidized subscriptions for other readers.

The success story of the *taz* is on one hand a story of innovative journalism, but it is also a story of active supporters – people who donate subscriptions for prisoners; doctors who display the newspaper in waiting rooms; readers who support the cooperative with a donation or simply join to become associates.

The non-profit-orientation of the *taz* is the base of the cooperative, making the newspaper different in comparison to many other high quality publications distributed by profit-oriented publishing houses. In this respect the model of the cooperative is not simply transferable to other media companies, as commercial enterprises will not receive such strong support from their readers.

The founding of the *taz* can be best compared to the self-organized and non-profit online portals which emerged in the years after the centennial. In 1978 it was evident that a newspaper could not live without paper, cost of printing, wages, etc. The fact that online publishing is practically free creates a huge problem when asking subscribers to pay for content. Traditionally only those who have a community based on solidarity are able to survive with an online offer based on professionally produced journalistic content.

Long before the Internet era, the *taz* made "friends" with their customers – first with their "rescue campaigns" and begging letters, and later by offering readers a chance to become members of the cooperative.

4 Relationship management in a digital age

As in every friendship, every so often the relationship with customers must be approved with a significant gesture. Therefore the *taz* has developed a very differentiated aftersales management, which is almost invisible from outside. Each year the *taz* sends at least two mailings and three newsletters to its 12,000 members, in addition to two mailings for another 5,000 interested people. Every day the cooperative receives approximately 100 e-mails featuring diverse questions, topics, and demands. The *taz* also carefully administers its data in order to send birthday greetings to its members.

The *taz* has roughly 7,500 e-mail-addresses from its associates and sends an e-mail to this list nearly twice a month in order to inform its audience about news from the *taz* including invitations to certain events. The newspaper organizes debates about current topics such as controversial advertisements in the *taz*.

The basis for what is called "customer loyalty" is an open and authentic communication, designed to win the devotion of readers, users, and associates – not only to ensure that customers think about their own cost-benefit-calculation, but also to create an emotional and personal relationship. The true effectiveness of "binding readers" has yet to be determined in the digital world, but a new campaign called "tazzahlich" (I pay for *taz*) is designed to emphasize its value and ultimately promote the concept.

In early 1994, *taz* became the first German national newspaper to post its complete print version on the Internet, free of charge. In 2007, *taz* began offering *taz.de*, which became its own constantly updated online version. While *taz's* complete print version is still available online for free, in May 2011 the newspaper began collecting donations by running a small tag below each article. The donation requests are meant to replace the "everything-for-free-culture" with a "culture of fairness," allowing users to decide whether to pay for one article or 100. Customers can choose to donate via Flattr or Paypal, or even by creating a *taz* account for express payment.

While the early stages of this campaign have been promising, what remains to be discovered is how to connect emotionally with readers on various social network sites, and whether this strategy will promote payment for online content. So far online relationships work differently:

–   The "like" button on Facebook is only a sign of affirmation but does not bear any sort of responsibility.
–   The follow-option on Twitter is a simple sender-recipient-model – one talks, the others applaud.

– The model of Google+, to be added to someone else's circle, is unthinkable in real life. Who would want to be added to a group without being asked permission?

It's not only a question of whether people like our journalistic products or if the payment methods are sufficiently barrier-free, it is also a matter of offering the most personal experience in hopes that a binding relationship is formed between the publisher and the customer. Ultimately, without intense and continuous contact, the relationship will not work.

Harry Browne
The promise and threat of foundation-funded journalism[1]

A critical examination of three case studies

Abstract: This paper considers philanthropic funding of journalism and looks at examples of journalistic institutions that receive prior funding (as opposed to post-facto reward) from charitable foundations. Drawing on sociological literature, it raises questions about the purposes of philanthropy, about the transparency of media that use philanthropically funded material, and about the assumption of a unitary "public interest" common to both philanthropy and to traditional journalism. It specifically examines ProPublica in the United States, Transitions Online in Eastern Europe and the Centre for Public Inquiry in Ireland. It concludes that both a critical understanding of foundations themselves and a consideration of the case studies presented should encourage wariness about philanthropic funding as an unproblematic model for the future of journalism.

Keywords: philanthropy, foundations, sociology, investigative journalism, journalistic bias

1 Introduction

With the idea that journalism is, in effect, a charity case having moved into the mainstream, a small but significant group of journalists and researchers have been examining how journalism has been, and might be in future, funded by charitable foundations. Concerns about bias and control, so prominent in consideration of state and commercial funding of journalism, have been somewhat lacking in the discussion of foundation funding.

Carol Guensberg's (2008) article on "nonprofit news" in the influential *American Journalism Review* set a tone of cautious hope, only slightly tempered by critical concerns, that has remained around the concept. The most oft-expressed worry has been that foundations will be forced to cut back on funding journalism

---

1 An earlier version of this article appeared as "Foundation-funded journalism: Reasons to be wary of charitable support", *Journalism Studies* 11, no. 6 (2010): 889–903.

because of their own financial worries in the wake of the global financial crisis. Westphal (2009) is among the researchers/journalists to highlight and welcome the support of foundations for journalism, with the major caveat being whether the charitable sector can do enough to make a real difference, and little attention paid to other, editorial, dangers potentially inherent in foundation funding.

McChesney and Nichols (2010), left-liberal critics of American mainstream media structure and bias, deal briefly with the issue in their study-cum-polemic on journalism's woes and the solutions to them. "Leaving aside the issue of whether we want foundations to have this much power," they write, "how realistic is the foundation-funding model for the next generation of journalists?" (McChesney/ Nichols 2010: 87) The authors – whose major concern is to encourage state support for journalism – really do leave that issue of power aside, concentrating instead on the cash caveat, i.e. how little money foundations have made available for nonprofit journalism: in 2008 it was "less than one-tenth of the annual newsroom budget of … *The New York Times*" (McChesney/Nichols 2010: 88). Having suggested that philanthropy is not equal to the scale of the problem in journalism – "we would feel a lot better if (the 20 million dollar paid to nonprofits by foundations in 2008) had a few more digits attached to it" (McChesney/Nichols 2010: 88) – they proceed nonetheless to praise the philanthropists: "we welcome foundations that want to write checks" (McChesney/Nichols 2010: 88) and to note that "there is much to celebrate in the willingness" of such foundations to support journalism (McChesney/Nichols 2010: 97). The recent trend toward foundation investment in for-profit companies, at least in the United States, may give further cause for such celebration (Strom 2011).

2 Looking a gift horse in the mouth

The present paper, based on three cases where journalism has been funded by foundations, and based also on fundamental questions about the role of foundations drawn from sociology rather than journalism studies, is essentially about looking a gift horse in the mouth.

It would be unfair to jump to critical conclusions via anecdote without more comprehensive analysis than is offered here. There is no doubt that there is some important work being done in and via the support of these institutions, as indeed the above-mentioned studies illustrate. Removing direct commercial pressures from some of the practice of journalism could, logically, result in an improvement in some of that journalism – by giving reporters more time to work on a story, by freeing them to pursue less-popular topics and by reducing the likelihood of pressure from an owner or advertiser.

The case study approach of the research presented here looks at some elements of the work of three significant foundation-funded journalistic non-profits, ProPublica in the United States, the Centre for Public Inquiry in Ireland, and Transitions Online in Eastern Europe, each organisation having been brought into being by a particular foundation. Clearly not every case of philanthropic support of journalism will involve such a close and organic relationship as existed in all three case studies here. A philanthropically funded journalistic organisation might have diverse funders, or an individual journalist might seek once-off financial support for a particular story.

The limited and specific case studies are preceded, below, by insights drawn from historical, sociological and political writings that raise more broad and basic questions about the role of foundations and, therefore, of their potential role in facilitating journalism that criticises and interrogates centres and structures of power. Combining these insights with the case studies, it is contended that "nonprofit news" raises some of the same problems as commercial journalism – including serving agendas that may possibly be hidden and hewing to establishment-defined ideological limits – while potentially adding some new ones of its own. As Russ-Mohl (2006: 200) has shown, even before the financial crisis, foundations such as Bertelsmann in Germany and Gannett's Freedom Forum in the U.S., despite their basis in media empires, have been idiosyncratic in their direct support for journalism, and for the education and research that might underpin it. However, even when support is fairly reliable, potential problems include: encouraging journalists to anticipate and chase after the whims of funders (some academics may be familiar with this phenomenon); creating awkward conflicts of interest due to the often-delicate relationships between charitable funders and the state bodies the journalists should be investigating; and subsidising the very news organisations whose conspicuous failures have helped to create the current crisis for the profession. (Davies (2008) has been joined by McChesney and Nichols (2010) as required reading for those seeking English-language analysis of precisely how those existing institutions are blame-worthy.) This article occupies itself principally with these three areas of potential objection to foundation-funding for journalism. An additional concern, not addressed in these pages but voiced by some practising journalists with whom the author has discussed this matter, is the possibility that foundation funding will push reporters towards "long-termism" and excessive seriousness and jargon in their work, moving the affected journalism further away from a mass audience as it becomes increasingly configured for foundation evaluators, policy-makers and other elites.

None of these issues should be regarded as reasons to dismiss foundations as potential sources of funding for journalism, which has never been pure and cannot afford to be choosy. But taken together they do suggest causes for concern that go beyond those voiced in the extant literature.

3 The benevolent fog

In a passage about the ethical confusion that may be engendered by foundations, Edmonds (2002) offers a basic note of caution about philanthropic funding of journalism:

> Here's a journalistic proposition: it would be ethical for a reporter to accept a grant from the Ford Foundation for coverage of Eastern Europe. (...) But it would be wrong to accept a grant from General Motors to cover international trade. GM's economic interests in the matter would create a perceived conflict of interest (...).

> Lost in the benevolent fog that surrounds most foundations is the notion that they may have more of an agenda, not less, than a sponsoring corporation. (Edmonds 2002)

Edmonds' example, contrasting attitudes toward funding from a foundation based on an automotive dynasty to funding from an automotive dynasty per se, is not purely theoretical: it is based, he writes, on the news-policy manual of America's National Public Radio, which makes precisely this distinction between foundation support (good) and corporate sponsorship (bad, at least potentially). The broadly skeptical thrust of Edmonds' research has had remarkably little echo in the years since it was published. Media analyst Jack Shafer, controversially fired from *Slate.com* in 2011, has been perhaps the most persistent and prominent critic of the foundation model (see for example Shafer 2009).

Bob Feldman (2007) is an exception in the academic literature. Writing from personal experience of left-leaning media organisations in the United States, he asserts – albeit largely anecdotally – that their politics have, broadly, been channeled in recent decades into "safe, legalistic, bureaucratic activities and mild reformism" (Feldman 2007: 427) largely through the influence of their foundation backers. He notes that those organisations that are "primarily concerned about threats to media independence (focus) all their attention (...) on for-profit or government control; they ignore the possible influence of large subventions from non-profit institutions such as foundations" (Feldman 2007: 428). Foundations often operate in a "climate of secrecy" (Feldman 2007: 428) and effectively manage the organisations they fund through meetings, conferences and suggestions, domesticating their agendas. To document the degree to which this sort of foundation support/management/pressure has resulted in left-leaning media turning safer and duller would "require a massive research project unlikely to find funding" (Feldman 2007: 429).

Feldman's tone of righteous indignation tempered by weary humour is common to the sociological literature that is more broadly critical of foundations:

> The critical study of foundations is not a subfield in any academic discipline; it is not even an organised interdisciplinary grouping. This, along with concerns about personal defunding, limits its output, especially as compared to that of the many well-endowed centres for the uncritical study of foundations. (Roelofs 2007: 387)

There are "more critical studies of foundation garments" (Roelofs 2007: 387), they write, than there are of foundations.

Concerns about the power and influence of foundations appear more likely to be voiced on America's conspiracist right, where George Soros in particular is a bête noir, than on the academic or political left. Marxist geographer David Harvey's attribution of some of the success of neoliberalism in recent decades to capitalists' "shaping of oppositional cultures through the promotion of NGOs" (Harvey 2009) is a typical passing but undeveloped echo of Feldman. Occasionally a specific foundation comes under critical scrutiny from the left, as when economist Rob Larson attacked the Clinton Foundation for being "funded by the people, governments, and companies that help create the problems that the charity seeks to address" (Larson 2009).

Research, and indeed polemic, from the underdeveloped realm of "critical foundation studies" has tended to focus on the effects of foundation funding on the priorities of academic researchers and global-development organisations. According to Arnove and Pinede (2007), basing their findings on long-term studies of the "big three" U.S.-based foundations – Ford, Rockefeller and Carnegie – "they have played the role of unofficial planning agencies for both a national American society and an increasingly interconnected world system with the United States at its center" with an "elitist, technocratic approach to social change" (Arnove/Pinede 2007: 392). They quote from a 1930 essay by Fabian theorist Harold Laski, who wrote:

> The foundations do not control, simply because, in the simple and direct sense of the word, there is no need for them to do so. They have only to indicate the immediate direction of their minds for the whole university world to discover that it always meant to gravitate swiftly to that angle of the intellectual compass. (Laski, in Arnove/Pinede 2007: 415)

Chasing after the mind of a proprietor or editor is not unknown in journalism. However, the supposition that the foundation represents a cleaner, less capricious form of direction than the commercial proprietor does not always stand up to scrutiny. "In 1996 and 1997 (…) the Ford Foundation (…) sent shock waves through the academic world by calling into question the validity of area studies programs that had been largely established and sustained by the Rockefeller Foundation and Ford Foundation" (Arnove/Pinede 2007: 414). The authors document how in the 1990s the foundations created bitter divisions in African and Eastern European academia.

Within the world of philanthropy it is not controversial that the activities of foundations are intended as an exercise of power for particular ends, though those ends are typically depicted as benign. Sean Stannard-Stockton, a columnist for the *Chronicle of Philanthropy*, has written of how philanthropists "attempt to shape events by providing or withdrawing grants"; he calls this "a form of hard power that leans heavily on the idea that influence is best achieved through offers of incentives or threats of penalties" (Stannard-Stockton 2010).

The central critique of foundations by critical scholars is more fundamental: that they are an important component of the establishment and maintenance of existing structures of elite control, both in particular states and within the larger global system. The extent to which, therefore, they can contribute to changing, or even scrutinising and critiquing, those structures must therefore be in some question. "We must continue to ask whether or not foundations can achieve an end that runs counter to the core interests of those who have contributed to create these foundations" (Fasenfest 2007: 382).

Foundations themselves are rarely held to account by journalists. One foundation president has acknowledged frankly that among the privileges enjoyed by foundations is the quiescence of the press:

> Foundations lack the three chastising disciplines of American life: the market test, which punishes or rewards financial performance; the ballot box, through which the numbskulls can be voted out of office; and the ministrations of an irreverent press biting at your heads every day. (Goldmark 1997)

4 ProPublica

In 2007 a charitable foundation, and effectively its single benefactor, created what is by its own account the largest investigative newsroom in the world, in the form of ProPublica. The New York-based non-profit organisation, directed by a former managing editor from the *Wall Street Journal*, Paul Steiger, is the creation of Herb Sandler, who with his wife Marion was boss of World Savings Bank. The couple were named in *Time Magazine* in February 2009 (Time 2009) as among the "25 people to blame for the financial crisis" for promoting "tricky home loans" with "misleading advertising." (No mention was made by *Time* of the Sandlers' munificence to journalism.) The Sandler Family Supporting Foundation, a funder of liberal causes (Nocera 2008) in the U.S., supports ProPublica with 10 million dollars annually. "Stories which have moral force, stories that are important to the sustainability of a democracy," Sandler, chair of ProPublica as well as its chief benefactor, said, "those are the stories I hope we will be doing" (Perry 2007).

Its provenance in the financially and politically active elite must raise questions about ProPublica, notwithstanding its particularly clear and comprehensive coverage of financial issues. ProPublica's first major report was a national/international story, produced jointly with the commercial news network *CBS* and its TV flagship *60 Minutes* programme. It was an investigation into another news organisation – the U.S.-government-funded Arabic TV station *Al Hurra*. The questions raised by the report go beyond the fact that, as Miner (2008) observed, it hardly filled a media void, given that the *Washington Post* did a similar exposé about *Al Hurra* on the same day.

The joint report (CBS News 2008, with material also available on *propublica. org*) carries the ProPublica logo but is otherwise difficult to distinguish from an ordinary *60 Minutes* report. It sets out to show that the U.S. government had been wasting its money by creating an Arabic news channel – and part of the report's method is to engage in borderline caricature of "dysfunctional" Arabs and to criticise the Virginia-based station for airing points of view, critical of Israel in particular, that are largely uncontroversial in the Arab world. The report certainly does nothing to challenge the common U.S. mainstream view that opposing Israel is inherently wrong; indeed it essentially and implicitly adopts that view. A revealing passage of the transcript includes an interview with an American who had been brought in to *Al Hurra* on what proved a futile mission to straighten out the "imported" Arab staff:

> Larry Register, a former *CNN* executive with 20 years of experience, who was brought in a-year-and-half ago to rescue the channel (...) says he found his staff of Arabs, imported from the region, divided along religious, ethnic and political lines.
>
> Asked what state the channel was in when he first walked in the Al Hurra newsroom, Register tells (60 Minutes reporter) Scott Pelley, "Dysfunctional, extremely dysfunctional."
>
> "Words like militias were thrown around," he explains. "There was this militia that was in charge of this, and this militia in charge of that."

> "It felt like you were living in the Middle East. It felt like somebody had picked up the Middle East and brought it to Springfield, Virginia, of all places," Register remembers.

> When Register wanted to put on breaking news his first week, he says he found his staff was out to lunch, literally. "There was nobody there. The whole newsroom was empty," he remembers. "Everybody'd gone to lunch. So I'm asking, 'Well, what is this?' 'Well, they take three hour lunches in between programs.'" (CBS News 2008)

No one notes that long breaks in the middle of the day, generally combined with late evenings, are standard working practice in the Mediterranean region. The "militia" comment, which could be interpreted as a suggestion that paramilitaries controlled various departments within the station, is left to rest as though it were a normal bit of Arab "colour," its significance unexplained.

Al Hurra, to be sure, could be legitimately criticised. A particularly egregious item on the Arabic station from a credulous reporter at an Iranian Holocaust-denial conference came in for appropriate opprobrium (CBS News 2008). But Scott Pelley's line of questioning to a station executive lumped it together with other aspects of the programming that would surely have enhanced its credibility among Arabs:

> There's a pattern here, critics of this channel say. You have Nasrallah (the Hezbollah leader) given an hour of air time. You have the Holocaust deniers conference covered. Now, you have this person saying that Israel is a racist state. Is this the kind of thing the American taxpayer should be paying for? (CBS News 2008)

It is arguable that for its first major report, ProPublica not only subsidised a massive corporate news operation, but that it did so within traditional American ideological constraints – most obviously the denigration of Arabs and almost-unqualified support for Israel.

5 Centre for Public Inquiry

In addition to its major funding from Sandler's foundation, ProPublica also receives some funding, albeit a relative drop in the ocean, from the Atlantic Philanthropies, the charitable foundation based on the fortune of Irish-American airport-duty-free entrepreneur Chuck Feeney.[2] Atlantic was the sole significant

---

2 The author was involved in a minor capacity in a project at his home institution in Ireland funded by Atlantic Philanthropies.

funder of the Centre for Public Inquiry (CPI), a short-lived Dublin-based investigative organisation run by one of Ireland's leading investigative journalists, Frank Connolly – whose reporting on political corruption, mainly in the planning process, had helped to bring about major state-run tribunals of investigation in the late 1990s (O'Clery 2007: 276). The brief year of operation of the CPI in 2005–06 tells a complex and cautionary tale about the nexus into which journalism enters when it forms relationships with the philanthropic sector.

The philanthropist behind Atlantic, Feeney, is famously shy. However, the respected veteran Irish journalist Conor O'Clery got considerable access to write a biography (O'Clery 2007) and O'Clery was subsequently involved in an Irish television documentary, a flattering portrait of the admirable and modest "secret billionaire" in May 2009 (RTE 2009). In that programme one interviewee intoned "I think he's a saint" and not need have feared any contradiction.

The book and programme, made with Feeney's cooperation, show that Feeney, through his quiet and conditional offers of cash from the late 1990s onward, effectively directed some higher-education policy in the Irish state and among other things brought about the creation of an allegedly state-directed funding initiative, the Programme for Research in Third Level Institutions. Whether this was a good thing is not a matter of concern here; the point is that, like many charitable foundations, Feeney's Atlantic Philanthropies was operating not simply in the NGO sector but in close cooperation with elements of the state itself. For example, in 2003 the foundation threatened the prime minister, Bertie Ahern, that Atlantic would stop paying for research in Ireland if the government insisted on cutting its own contribution: Ahern obliged by using private pressure and press leaks to force the hesitant education minister to maintain state support for the sector (O'Clery 2007: 274 f.).

Feeney had met journalist Frank Connolly during the 1990s in the course of the billionaire's involvement, together with other Irish-American business people, in the Northern Ireland peace process (O'Clery 2007: 276 f.). After several friendly meetings they came to discuss Connolly's work on political corruption, and Feeney told Connolly that Atlantic had helped to fund an investigative body, the Center for Public Integrity, in the United States. By 2004 Connolly and Atlantic Philanthropies had developed a plan to establish an analogous body in Ireland (Connolly, interview with the author, 23rd February 2009). "Connolly, a serious, methodical investigator, seemed an ideal choice" as director (O'Clery 2007: 276). The CPI would get four million euros funding for its first five years of work, beginning in 2005. Former Irish prime minister Bertie Ahern (himself later the subject of investigations, including by Connolly, that forced him out of office in 2008) told the documentary-makers that, when he heard of this plan to finance such an organisation, he approached Feeney directly to tell him that it was not necessary or advisable (RTE 2009).

The matter was complicated by the fact that the CPI director, Connolly, was known for his left-leaning views and investigative pursuit of Ahern and other senior political figures. Furthermore, Connolly had family ties to the IRA – his brother Niall had been arrested in Colombia in 2001, allegedly making contact with rebel groups there. Strong criticism of Connolly and the CPI was voiced publicly by politicians, and some journalists, especially in Tony O'Reilly's Independent group of newspapers, took up the campaign against the CPI (O'Clery 2007: 277 ff.).

The centre's first two investigative reports were published in the second half of 2005 in what were intended to be the first two editions of a new publication, *Fiosrú* ("enquiry" in the Irish language). They were generally seen as scrupulous and well respected studies of, first, conflicts of interest in planning around a historic site in Trim, County Meath, and, second, the complex legal and political history of a controversial Shell gas-pipeline project in County Mayo. The latter, in particular, was a strong intervention in a major public dispute that had seen (and has continued to see) hundreds of police dispatched to a remote coastal location in the west of Ireland, and the arrest and imprisonment of a number of protesters. The CPI report came carefully down on the side of the protesters against Shell, the government and the pipeline, and raised questions about the political and planning decisions in the background to the project and in relation to other deals for oil and gas exploration off the Irish coast (Providence Resources, an oil and gas exploration company, is controlled by the same O'Reilly family that dominates the Irish newspaper industry) (Centre for Public Inquiry 2005a).

The next CPI investigation intended to probe the Dublin Docklands Development Authority, where politics, finance and property-development intersected – like the first two reports, the sort of story that needs a lot of time and context, the resources that "ordinary" journalism finds itself largely unable to provide. The CPI's five-year plan was, according to Connolly, an ambitious programme that would have taken it to the highest levels of the political establishment (Connolly, interview with the author, 23rd February 2009).

At this point, late in 2005, the Minister for Justice Michael McDowell, by his own public admission, leaked to a well-known journalist for Tony O'Reilly's *Irish Independent* newspaper some documents from an investigation into Frank Connolly that appeared to suggest Connolly had several years earlier given false details in a passport application in order to travel to Colombia. Connolly made a public statement on 7th December 2005:

> The Minister has sought to interfere with, if not jeopardise my employment as Executive Director of the Centre for Public Inquiry. By disclosing confidential information from Garda

files to a member of the board of Atlantic Philanthropies, which funds the CPI, which is clearly insufficient to support a prosecution against me, he has intended to damage my reputation and my career as an investigative journalist. (...)

Furthermore, confidential documents from a Garda investigation file were copied to Independent Newspapers to the damage of a citizen, who is entitled to the presumption of his innocence and to the protection of his good name. (Centre for Public Inquiry 2005b)

The allegations against Connolly were never proven; however, the now wide-open hostility between Connolly's CPI and the Irish government was causing discomfort among Atlantic's representatives in Dublin – who had to work with state bodies in relation to other projects – and through them at Atlantic's headquarters in New York (Connolly, interview with the author, 23rd February 2009). In December 2005, in an answer to a parliamentary question, McDowell (under parliamentary privilege) tied Connolly's alleged activities to the Colombian rebel FARC organization and to narco-terrorism. At an Atlantic board meeting in New York, a fax arrived from Dublin containing McDowell's charges: after reading it, the board decided that the foundation could no longer fund CPI while Connolly was in charge (O'Clery 2007: 283). Connolly, however, would not step down and the CPI's own board of directors (comprising a senior journalist, a lawyer, a theologian and a former High Court judge) released a statement to the press expressing support for Connolly (Centre for Public Inquiry 2005c).

Atlantic nonetheless withdrew funding and the CPI was out of money and therefore, within a few weeks, no longer able to operate (O'Clery 2007: 283 ff.). Several years later, its brief history remains open to debate; however, for the purposes of this study it is relevant that Atlantic Philanthropies abandoned its funding of an investigative-journalism organisation because of sensitivity about the relationship between its director and the government – or, by the very best interpretation, because, encouraged by the government, it came to negative conclusions about that director's character and behaviour without due process. (Connolly is today press officer for Ireland's largest trade union.)

Atlantic has since gone on to support the Huffington Post Investigative Fund, causing McChesney and Nichols to praise it as "a journalism-oriented, highly engaged foundation" (McChesney/Nichols 2010: 97).

6 Transitions Online

The final case study is more briefly considered and lacks the obvious political and journalistic drama of the first two. It relates not to high-profile investigations but to a long-term project by a major funder to influence the development of Eastern

European societies in the post-communist period. Transitions Online (TOL) is a partly Web-based NGO project, centred in the Czech Republic, established in 1999 as a successor to *Transitions* magazine. The "transition" in the title refers to Eastern Europe's 28 post-communist states, which TOL covers. Its own work is largely in English, though it offers training to East European journalists who work in their own languages. The original magazine was aimed largely at an academic readership and TOL continues to specialize in education issues (Druker, interview with the author, 29th January 2009).

Seeded by George Soros's Open Society Institute, the TOL project was part of its funder's wider project to influence the intellectual direction of the region over the last two decades. It can be argued, of course, that the Soros influence on the region has relatively benign; it cannot be plausibly argued that his influence was neutral in terms of the desired outcomes in political and economic policy. Guilhot (2007) has studied how Soros set out to effect policy favourable for his business interests in the post-communist states by supporting "transitional" academic projects and creating, in 1991, the Budapest-based Central European University.

> Philanthropic practices allow the dominant classes to generate knowledge about society and regulatory prescriptions, in particular by promoting the development of the social sciences (...). Philanthropy offers a privileged strategy for generating new forms of "policy knowledge" convergent with the interests of their promoters (...). Far from seeking to curb the excesses of economic globalization, such efforts are actually institutionalizing it by laying the foundations of its own regulatory order. (Guilhot 2007: 447)

Such a project has a role for journalism. Leslie Sklair writes of how it would be directed not only at political and intellectual elites in the region but at a wider public there and elsewhere: Soros and other "corporate philanthropists ... embody the public relations thrust of the new globalising (transnational capitalist class)" (Sklair 2007: 26).

At the time the present research was conducted, TOL was no longer funded entirely by Soros. As donor interest has tended to move east across the former Soviet Union, TOL has received a mix of foundation funding (including continuing Soros money for its work on education issues) and support from state-based bodies such as the American National Endowment for Democracy and the Czech foreign ministry for its work on "democracy promotion" in Eastern Europe and southwest Asia. TOL's director Jeremy Druker said in an interview that funders don't interfere with the NGO's activities but explained that fundraising does involve persuading donors that "we share your values" (Druker, interview with the author, 29th January 2009). The organisation also generates income from "training," often involving Western journalists who come east to teach classes to aspiring journalists. In a further example of foundation-supported journalism subsidising traditional commercial operators, TOL's network of (mostly young) journalists has provided coverage of Eastern Europe for the U.S. magazine *Business Week*, which no longer directly supports a group of independent "stringers" there. According to Druker, the relationship with *Business Week* (which does not disclose to its readers the ultimate source of this coverage) is "non-commercial" but good for the organisation (Druker, interview with the author, 29th January 2009).

Thus TOL is a journalistic NGO providing business-friendly coverage of Eastern Europe with funding from an investor with enormous interests there, and incidentally in the process subsidising a commercial news provider that reaches a large American audience.

Such a lack of transparency is commonplace when other mainstream journalism providers use material generated from these not-for-profit outlets. In November 2009, for example, *The New York Times* published a story by freelance journalist Lindsey Hoshaw, about a Pacific Ocean "garbage patch." At the end of the story was this simple note: "Travel expenses were paid in part by readers of *Spot.us*, a nonprofit Web project that supports freelance journalists" (Hoshaw 2009). The implication, picked up by the wire agency AFP (2009) and reported widely online, was that the article was simply "crowd-funded," commissioned through the enthusiasm of hundreds of donors. Nowhere in the original article or in the AFP report was it pointed out that *Spot.us* came into existence thanks to a grant from the Knight Foundation.

It is arguable that the appearance of foundation-supported material, without clear indication of its ultimate financial provenance, should be regarded as insidious, in much the same way that so many studies of journalism view the proliferation of PR-generated material.

7 Conclusion

The increasing role of direct foundation funding for journalism might nonetheless be a cause for celebration, if there were strong reason to believe that the ultimate source of subsidy was both (1) always clear to readers and (2) democratic and responsive to the wider public. However, on examining the cases outlined above and considering the arguments about the nature of foundations themselves, there is at least some reason for concern as to whether these conditions can be met, or whether such support brings new worries for the credibility and viability of journalistic institutions.

It also raises a number of theoretical concerns of interest to journalism scholars. If, as van Dijk (2009) suggests, news can be regarded as form of ideological discourse, how does foundation support affect both the "social knowledge"
(van Dijk 2009: 195) and the immediate context of journalistic participants? If, as
Guilhot (2007) suggests, foundations have explicitly ideological programmes to
generate such social knowledge, by what means would this become manifest in
reporting, and how might such manifestations be detected by a researcher?

Many journalistic professionals perhaps welcome foundation support
because, in the context of rapid change in journalism practice driven by technology and finance, it appears to offer them a return to what has been called the
"high modernist" conception of "professionalism," with journalists "conceiving
of themselves as, in effect, a representative or stand-in for a unitary but inactive
public" (Hallin 2000: 234 f.). The idea of such a unitary, passively constructed
"public interest" is central to the discourses both of traditional journalism and of
the philanthropic sector. However, a serious analysis will be forced to admit the
possibility that interests come in many shapes and sizes, and operate on all sorts
of potentially competing and hidden agendas.

References

AFP. New York Times publishes "crowd-funded" article. Retrieved on 13th November 2009,
    from http://www.google.com/hostednews/afp/article/ALeqM5iSC_k8BSFHIGN6n-
    bX0mpFbUaF1Tw (2009).
Arnove, Robert/Pinede, Nadine. Revisiting the "big three" foundations. *Critical Sociology* 33
    (2007): 389–425.
CBS News. US Funded Arab TV's credibility crisis. 60 Minutes. Retrieved on 12th May 2009, from
    http://www.cbsnews.com/stories/2008/06/19/60minutes/main4196477.shtml (2008).
Centre For Public Inquiry. The great corrib gas controversy. Fiosrú 1, no. 2, 2005a.
Centre For Public Inquiry. Press release 16th December 2005. Retrieved on 7th August 2009,
    from http://www.publicinquiry.ie/pressreleases.php (2005b).
Centre For Public Inquiry. Statement of Frank Connolly. 7th December 2005. Retrieved on 7th
    August 2009, from http://www.publicinquiry.ie/pressreleases.php (2005c).
Davies, Nick. *Flat Earth News*. London: Chatto & Windus, 2008.
Edmonds, Rick. Getting behind the media: What are the subtle trade-offs for foundation-funded
    journalism? Retrieved on 12th May 2009, from http://www.philanthropyroundtable.org/
    article.asp?article=1104&paper=1&cat=147 (2002).
Fasenfest, David. Notes from the editor. *Critical Sociology* 33 (2007): 381–382.
Feldman, Bob. Report from the field: left media and left think tanks – foundation-managed
    Protest? *Critical Sociology* 33 (2007): 427–446.

Goldmark, Peter. President's letter. In annual report of the Rockefeller Foundation. Retrieved on 6th August 2009, from www.rockfound.org/library/annual_reports/1990-1999/1997.pdf (1997).

Guensburg, Carol. Nonprofit news: as news organizations continue to cut back, investigative and enterprise journalism funded by foundations and the like is coming to the fore. *American Journalism Review* 30, no. 1 (2008): 26–33.

Guilhot, Nicholas. Reforming the world: George Soros, global capitalism and the philanthropic management of the social sciences. *Critical Sociology* 33 (2007): 447–479.

Hallin, Daniel C. Commercialism and professionalism in the American news media. In *Mass Media and Society*, Curran, James/Gurevitch, Michael (eds.), 218–237. London: Arnold, 2000.

Harvey, David. Organizing for the anti-capitalist transition. Talk for the World Social Forum 2010, Porto Alegre, Brazil. Retrieved on 18th March 2010, from http://davidharvey. org/2009/12/organizing-for-the-anti-capitalist-transition/#more-376 (2009).

Hoshaw, Lindsey. A float in the Ocean. Expanding islands of trash. Retrieved on 13th November 2009, from http://www.nytimes.com/2009/11/10/science/10patch.html?_ r=1&scp=1&sq=lindsey%20hoshaw&st=cse (2009).

Larson, Rob. The Clinton Foundation Donors. Retrieved on 18th March 2010, from http://www. counterpunch.org/larson01282009.html (2009).

McChesney, Robert W./Nichols, John. *The Death and Life of American Journalism: The Media Revolution That Will Begin the World Again*. Philadelphia: Nation Books, 2010.

Miner, Michael. Is Pro Publica living up to its promise? Retrieved on 14th May 2009, from http:// www.chicagoreader.com/TheBlog/archives/2008/07/07/pro-publica-living-its-promise (2008).

Nocera, Joe. Self-made philanthropists. Retrieved on 8th December 2009, from http://www. nytimes.com/2008/03/09/magazine/09Sandlers-t.html (2008).

O'Clery, Conor. *The Billionaire Who Wasn't: How Chuck Feeney Secretly Made and Gave Away a Fortune*. New York: Public Affairs, 2007.

Perry, Suzanne. Financier backs project to beef up investigative reporting. Retrieved on 15th May 2009, from http://philanthropy.com/free/articles/v20/i02/02001001.htm (2007).

Roelofs, David. Note on this special issue of Critical Sociology. *Critical Sociology* 33 (2007): 387–388.

RTE. Secret Billionaire: The Chuck Feeney story (TV documentary broadcast on 5th May 2009).

Russ-Mohl, Stephan. The economics of journalism and the challenge to improve journalism quality. A research manifesto. *Studies in Communication Sciences 6*, no. 2 (2006): 189–208.

Shafer, Jack. Nonprofit journalism comes at a cost. Retrieved on 13th November 2009, from http://www.slate.com/id/2231009/ (2009).

Sklair, Leslie. Achilles has two heels: Crises of capitalist globalization. In *Thinker, Faker, Spinner, Spy: Corporate PR and the Assault on Democracy,* Dinan, William/Miller, David (eds.), 21–32. London: Pluto Press, 2007.

Stannard-Stockton, Sean. Philanthropists' "soft power" may trump the hard pull of purse strings. Retrieved on 22nd April 2010, from http://philanthropy.com/article/Soft-Power-Could-Be-More/65080/ (2010).

Strom, Stephanie. To advance their cause, foundations buy stocks. Retrieved on 2nd December 2011, from http://www.nytimes.com/2011/11/25/business/foundations-come-to-the-aid-of-companies.html?_r=1&emc=tnt&tntemail0=y (2011).

Time. 25 people to blame for the financial crisis. Retrieved on 14th May 2009, from http://www.
      time.com/time/specials/packages/article/0,28804,1877351_1877350_1877343,00.html
      (2009).
van Dijk, Tuen A. News, discourse and ideology. In *The Handbook of Journalism Studies*,
      Wahl-Jorgensen, Karin/Hanitzsch, Thomas (eds.), 191–204. London: Routledge, 2009.
Westphal, David. Philanthropic foundations: Growing funders of the news. USC Annenberg
      School for communication. Center on communication leadership & Policy research series.
      Retrieved on 7th August 2009, from http://communicationleadershipblog.uscannenberg.
      org/Westphal-Philanthropic%20Support%20for%20News%20report.pdf (2009).

Joachim Meinhold

Resisting crisis rhetoric in financing journalism

A managerial and customer oriented standpoint

Abstract: At least when it comes to the German market we argue that the "crisis rhetoric" about newspapers is highly questionable. The traditional financing model for journalistic content, enhanced by new but related businesses, still enables a return on sales well into double-digit figures as it is the case in well managed newspaper publishing houses. The model does indeed require a critical discussion on journalistic and publishing optimization in terms of strategic perspective. In this context we discuss ten issues which should be approached to optimize the newspaper business. Alternative financing models like non-profit organizations and cooperatives on the other hand are seen with scepticism. They especially bear dangers concerning the professional governance of publishing houses and the preservation of editorial independence.

Keywords: business model, advertising sales management, customer relation management, cost structure, industrial organization of newspaper publishing houses

---

1 Crisis rhetoric

The subject under discussion is the financing of journalistic content, which is deemed to be in crisis. One assumes a need for new business models. If one follows this "crisis rhetoric" in the media (and within parts of the scientific discussions), one is reminded of the American saying: "They make so much noise that they believe their own bullshit." Is there really a reason in Germany for this hectic debate about, rather than with, the newspapers, or is it just about strategic concerns, challenges and responsibilities?

My considerations are mainly based on the home markets of the *Saarbrücker Zeitung Publishing Group*, where I am CEO, and the market of Berlin, where I have practical experience.

In 2011, *Saarbrücker Zeitung* is celebrating its 250th anniversary. Today our media, publishing, documentation and translation group has sales of approximately 350 million euros p.a. Our three main newspaper brands *Saarbrücker*

Zeitung, Trierischer Volksfreund and *Lausitzer Rundschau* have a sold circulation of 335,000 copies; more than 90 percent of them are subscriptions. In addition to regional newspaper publishing we are engaged in free newspapers, which are delivered to households, local radio, online portals, postal delivery, telephone directories and via our subsidiary *euroscript S.A.* in document management and translation services.

In fact, the debate about newspapers has been sceptical since their inception: Whether it be Gretchen in Dr. Faustus, who insists on learning of the decease of her beloved from the pejoratively named Blättchen (engl. "journal"), Balzac writing a novel about "Lost Illusions," or Karl Kraus asserting that only what is written between the lines "is not paid for"; thus we have never had a truly "good press." Indeed it is hard to perceive how we have managed for so long amid so much incompetence. Today too, our media journalists engage predominantly with our demise. Never has an industry been so "talked down" with such relish by its own people. Facts themselves have little impact, and indeed even the optimists among publishers pay them scant attention.

As far as I can see, it makes little sense to constantly declare one's own business model to be dead, only to have a potentially long wait for its demise. Furthermore, the newspapers are running their business under very different conditions: Not every crisis of a singular newspaper stands for a crisis of the branch of business with its traditional model of financing editorial content.

2 Facts

The average return on sales for our media, publishing, documentation and translation media house is reassuringly in the realm of double digits, and indeed at a level which the proud German export industry, mechanical engineering, has not achieved for many years. There are, of course, less successful papers, some are successful too, but I don't know better ones. In this situation I feel a little bit ashamed to talk about public subsidies.

Hard circulation figures in our very loosely structured and demographically complex home markets fell year on year in two locations by a delta of approximately one percent and in the area of Lusatia, which is located in the eastern part of Germany, on historical grounds and due to management errors, by just under three percent. *Saarbrücker Zeitung* and *Trierischer Volksfreund* are significantly better off than the average of German newspapers in terms of circulation development, and *Lausitzer Rundschau* stands out above the average of the "East." In all our subsidiaries we diminished circulation losses in the last three years.

Our reach in the regions we cover is quite high and even more stable than circulation. For instance, of just under one million inhabitants of Saarland, which is one of the 16 states in Germany, 460,000 over 14 years of age, according to media analysis, regularly read the *Saarbrücker Zeitung (SZ)*, and of two-person households in Saarland, over 60 percent have an *SZ* subscription or buy it at newsagents. However, we perceive the growth potential to be small in a context of increasing competition within the media business, though we are holding up well. No other medium can show comparable recipient range and quality. Nonetheless, there is a need for improvement in many respects. Of course, we are no longer in a non-competitive monopolistic position, but we are far ahead of others with our production of regional and local editorial content. The digital and print competitors do not catch up with the credibility of our brand and our editorial regional content.

Naturally, in the advertising business, we – like the whole newspaper industry in Germany – suffer from advertising-placement decisions of six important trade chains like Aldi, Schlecker or Lidl. The old "one third circulation – two thirds advertising" rule concerning the financing of editorial content has been transferred in recent years into a "50:50" rule. Financing from reader markets is increasing and this process will continue. Copy price increases which are part of this process are not helpful as regards coverage in recipient markets. But newspapers were always a product for that segment which stands out above the average on the basis of income, education, age and professional status.

We are certainly not pleased with the advertising-placement decisions of the trade chains against newspapers and in favour of direct distribution to home letterboxes or indeed postal services or Internet. In terms of advertising strategy, we also consider such an approach to be inappropriate, but clearly we will get through it.

In addition, in all locations the local advertising, classified and insert business, apart from trade chains and discounters is achieving positive growth rates of between two and six percent in 2011. In one, perhaps two locations, overall growth is even significantly up on the previous year, even in a context of increasing competition within the media industry.

To date, we are far from having exhausted the potential local and regional business in the traditional model. An overall assessment becomes even clearer if one takes account of the fact that substantial parts of previous (commercial) advertising business today (admittedly with lower contribution margins) remain "lodged" with our advertising journals, which are delivered for free, putting a range of 750,000 readers in the Saarland at our customers disposal.

This too finances editorial copy, which alas falls outside the remit of discussion, with its fixation on "newspaper content."

Thus there is no change to our strategically cautiously optimistic overall assessment, even though, as regional newspapers, we have already been largely excluded from the national advertising business in the past. After all, we have "survived" that.

Finally: The question is not about "content financing in general," but the financing of expensive publishers and editorial departments with commensurate wages! We strive to pay our journalists a commensurate wage; the average cost of one statistical editorial team employee (including all social-insurance costs, not including the costs of chief editors) in Saarbrücken is over 85,000 euros. If one reads about the business models of some leading bloggers in the U.S., clearly they aim to enter a kind of "cheap journalism" in every respect. One learns nothing of regularity of news production, analytical depth, time-consuming research, to name but a few. However, the salaries of journalists and platform operators in those business plans fall far below the level that is usual here in the non-wage-tariff-linked advertising-journal sector. We do not wish to become, and indeed we must not become, merchants of fear; we pay a tariff-based wage (which, in the structure existing, we do not consider a good thing).

However, neither in the past nor for the foreseeable future do we need to position ourselves as a "cultural asset" or "endangered species," as Mr. Kilz of the *Süddeutsche Zeitung* once proposed, but, if our business model sometimes does not run entirely smoothly, we should first remember a simple principle: "Do your homework first … ". Let me point out ten of those to-dos:

1) "Would you go shopping there?" is a question sometimes posed anxiously to friends. If one reads our ad price lists, one is overcome by a feeling as if one is contemplating the purchase of a bottle of red wine from a government "liquor store" in Finland in the 1960s, so immense is the prohibitionary impact. It is impossible to understand the sense of it all; no IT system can meaningfully illustrate this complexity; costs running into millions arise annually with SAP programming, merely because our price lists and ad managers are the way they are. The price differences are so vague, incomprehensible and unreal, that only experts can put together a reasonable package for customers or agencies. Without a personal presence, it is no longer at all possible to offset the negative, prohibitionary impact of our price lists in a business development discussion. And all this happens in the context of young agency staff that does not have a particular affinity with the press.

2) Many newspaper publishers are largely dependent on their major customers or key accounts. Why is this? Generations of managers in ad departments have become used – in the "hard-fought battle" with "their" key accounts – to "safe customers" at high prices, and the regional markets have been

neglected. The "Special Bike Shops Publication" to mark a particular occasion, which formed part of hand-to-hand sales wars, was at times too little for us and is today "too much" for our sales staff. In Germany, selling has a slight taint of decay.

3) Average sales at the newspaper publishers are far too high per order. Advertising journals have the capacity to show us how to build up volume. For instance, with sales of 0.8 million euros p.a., many small businesses just manage to achieve positive results and perhaps feed a family of three. For them too, we need to structure offers, even if they do not advertise based on a budget running into millions. The long-tail policy applies also in the print advertising business!

4) As for the trade, the advertising-placement decision in favour of direct distribution versus newspapers is generally not "Internet-related" but due to the interest in a 100 percent household coverage. It is possible that neither does the journalistic spread of a regional newspaper fit any more, and perhaps we will also have to reflect more on attractively combined offers of newspapers and other print titles. While, in specific areas, the paper lacks attractive household coverage, we must indeed structure it using supplementary distributed products. After all, the management and infrastructure are present. And after all, local magazines cannot only be produced by parish newsletters.

5) The differentiation criterion as between successful and less successful publishers consists in our group amongst other factors of the number of special publications. Those which have rising advertising sales produce one or two special publications daily, and cover almost 100 percent of the events eligible for special publications, whilst others have fewer special publications and overall lower advertising sales.

6) Even though most publishers have an excess of IT facilities, only very few have a functional, sales-oriented CRM tool enabling regular contact with potential customers of the region. Of our 58,537 businesses/organizations in Saarland identified in CRM, over the past twelve months, only about approx. eleven percent have been active customers of Saarbrücker Zeitung, and we have either not or indeed never approached most of the rest. At our other locations, the situation is not much better. Up to now the new business opportunities related to CRM have not been considered.

7) If one can buy an identical stereo system in one place for 5,000 euros and in another for 500 euros, then things will soon go wrong. God only knows why we have such appalling price differences among our cost-per-thousand prices not only in nominal terms, but also real terms, and not only in the sector, but also within newspaper groups. "Efficiency of markets" should also relate to transaction and information costs. It dates back to a time when editorial

fees were on average leveraged high above those of assistant professors and the cost burden was passed on as inventively as possible to the price lists. Naturally, I am convinced that our prices are entirely justified, but we will have to face the challenge, also in editorial and publishing organizations, arising from the fact that there are more advertising media than before, and cost-per-thousand prices are at least tending to fall. The marginal benefits of the last advertising euro invested in the media markets will balance out. Why the laws of Gossen of balancing out marginal returns should not apply in the media business? We are suffering more from the process of rebalancing the real price level in the advertising business than from a fundamental decision against print.

8) The publishers are not yet willing to stoop so low. Most just watch fairly apathetically when Google advertising vouchers worth 50 euros are sent to addresses registered with the Chambers of Commerce and Industry in the regions and offer medium-sized enterprises, lawyers, doctors, etc. extremely cheap entry into the Adwords option. If, as a managing director, one asks a sales manager to offer as an option a combination of Adwords, Linktipp and print advertising at a volume of approx. 500 euros with a guaranteed number of clicks on the homepage of the potential customer, in order to continue "monopolizing" customer access, one's interlocutor enumerates elaborately and with relish how, in view of the high (staff) business development costs incurred by the publisher, this is not worth it. The important task is on the contrary to concentrate on the relevant regional key accounts! In this way, entire customer realms are lost and regional entrenchment is reduced, a negative trend also from a journalistic point of view. For ads "sell" in reader markets too. There are still too few refined hybrid advertising offers, e.g. products for the job market or home improvement market. It will also depend to some degree not only on "having everything," but also on "how to combine it."

9) The "National Ads Question" or "at a distance, everything becomes poetry" (Novalis): Regional newspapers in Germany are weak in terms of national advertising. Of course, it is a seductive thought to think of re-positioning the national sales organizations of all newspaper publishers, for instance like *Publicitas* in Switzerland, and to offer the agencies solid, structured insertion units with reasonable prices, so that the newspapers do not get left out of national advertising in favour of TV and the Internet and build an appropriate counterweight against the increasing power of demand of media agencies. As appropriate as those thoughts concerning the national ads question are, so far, alas they have remained out of date and without consequence. The result of those 15 years old thoughts was the national ad initiative *NBRZ*, which is successful, but not sufficient for the future. Unlike others, I do not

know what volume can be gained through the project, which we should of course support strongly on one hand. But I have to concentrate on the other hand on those things which I can change and structure on the horizon of my own life, and not to focus on a cloud cuckoo land which can then not be built due to prevailing vested interests, and which only creates lucrative consultancy jobs. Naturally, it is sad that especially papers like ours are largely excluded from national advertising with regions around Cottbus, Trier and Saarbrücken. The advertising mainstream now runs from Munich via Frankfurt and Düsseldorf to Hamburg. This will not change quickly. Moreover, I have some doubts as to whether "hybrid advertising marketing" comprising print, online and radio is at all realistic on a centralized basis. Our experience with *OMS*, Germany's national online sales organisation, has not been encouraging so far; our own regional online advertising and our sales via cross-media regional packages are growing significantly faster than the *OMS* marketing. Thus, for this reason, I believe that we should first do our homework in the regions and put more focus on attractive made-to-measure advertising solutions for regional partners.

10) We still have baroque production structures, which demand to be financed. Many of us do not focus the issue of industrial organisation ("Verlagsfabrik") between corporations especially owned by different shareholders. Shared services, more scale, redefining of all (including the editorial) workflows, more focus on the market segments, those issues are not in the centre of managerial considerations. Newspapers get soon exhausted by the challenges of innovation and sometimes invest comparably little in new adjoining business areas. At the same time some newspapers decorate their profit-loss-accounts with the interests to be paid for their nice skyscrapers giving the impression, they were multinational players or real estate investors and not regional media publishers.

Overall, using our present business model for financing editorial content, we will have to accept new real price levels, and to believe anything different would be illusory. However, we will also be able to assign prices to unique quantitative and qualitative selling features of the print sector, in future too, although perhaps this will no longer be enough to finance the excessive proliferation of our content and publishing processes in the past.

3 High Internet growth rates

The advertising medium of the Internet is currently growing steadily at a percentage rate running into double digits, but what is the achievement rate of those ads? I can no longer bear to read and hear such reports on percentage increases without any mention of the starting level. Our *Saarbrücker Zeitung* editorial content costs just under 13 million euros annually, which is the price for six days of reporting with relevance to the social, political and cultural life of Saarland, as such it is a focal point of the region's public life. I don't know an Internet site being able to finance this amount in Germany every year and I am in some doubt about the possibility of much lower production costs.

Certainly there is much that can be optimized of our costly produced content, precisely in view of the Internet competition, and particularly with the younger and older generations, but I would say this: This brand, *Saarbrücker Zeitung*, which has stood for serious regional journalistic content and a broadly accepted selection of content for over 250 years, cannot over the medium term be displaced by blogs, Internet-supported citizen journalism, or social media. Thus people will continue wanting to have their information confirmed in the "paper" or "*Blättchen*" (or possibly in addition via the Internet or iPad) and they will insist on the bundle of content with its credibility provident by the regional newspaper. This coverage will also remain relevant in terms of advertising. But it also costs money to produce serious journalism.

To date, the Internet cannot demonstrate such advertising-impact analyses as can the publishers with their media analyses, various reader analyses or analyses from *ZMG*, Germany's central newspaper marketing research institution. And not without reason: Internet advertising is "cool," a "bargain," with all the leading technical facilities from tracking to interaction. Many advertisers are however not content with page impressions and unique users of photo galleries, game portals etc.; they want to have an environment comprising serious editorial – and local – news content.

I do not intend here to relativize the immense potential of this medium, and neither am I arguing that we should simply do some neat calculations by adding together, for instance, our total print and online reach. Certainly things cannot be done so simply.

It seems to me that there are high thresholds which make it difficult for other (digital) competitors to enter the regional markets of regular editorial content, first, because there exists no finance for the newcomer, and, second, because many of the newspapers have a deeply rooted defence system in their home markets.

4 Are we as good as we think we are?

Coverage and loss of circulation – these do not decline "just like that" or because of expressed product satisfaction. If we really had more regional journalistic quality, we would also be more unique in terms of attracting reader interest. How the truth really is, is something everyone can evaluate honestly for himself. It seems to me that, journalistically, mediocrity prevails on a daily basis, and one does not always uncover that daily "gold nugget," which readers are happy to pay for.

While our editorial teams are content within themselves, our readers are however not always so (e.g. younger readers and other segments as "liberal performers").

Moreover, according to our cohort analyses, the difficult aspect is that we are in particular losing our coverage among those who were aged around 30 in 1995, when the Internet did not yet exist. And so it would be appropriate to reflect for once in a self-critical way on the editorial and publishing strategies that led to this. Moreover, the disturbing side of this trend is that the potential losses of coverage in favour of the Internet largely still await us! The demands in journalistic terms will grow on condition of fierce competition of other media.

Within the "daily business" we dedicate little time to the topic "quality control of editorial content." There is little willingness for critical self reflection. If a steel plant delivers not properly produced coated sheets, the customer returns them instantly. Our readers sometimes just suffer quietly and give us up in time.

5 Don't worry, be appy …

We do indeed have experience in the digital arena; our online services are in profit on balance (without any hocus-pocus), and performance in respect of our regional transaction and advertising business is headed on a clear upward trajectory. We have online shops selling wine, selling tickets, supplying third parties with event data or editorial content for charge. Our writers are active on Facebook etc., and we post our contributions too.

From the beginning of 2012, we will be experimenting with "soft paywalls," through which revenue can be obtained for content from heavy users without the effect of "fading" traffic.

All of our newspaper titles, radio channels or sport magazines have priced iPhone and iPad applications, and we also produce and run them for third parties, e.g. the city information system *saarbruecken.de* with its related app.

These are all important studies and exercises for the digital future. We want to be editorial experts on all distribution channels. We can also envisage expanding our shop systems and to chargeable Sunday publications.

The only thing is that editorial departments where the average salaries exceed the level of assistant professors at universities are not sustainable. What is marketed on the Internet or via apps in terms of content largely originates free of charge from the domain of print media. The chargeable retrieval of content is limited, to put it mildly. Now studies looking to the future never sound very appealing at the outset, but this is no reason to omit them. However, the extent of academic and media attention to the subject of "apps" appears to me disproportionate, while issues such as the reorganization – and reanimation – of our sales departments, training for new sales agents, practicable CRM tools, cross-media standard projects and last but not least questions concerning the content which is to be paid for, should attract far greater attention. First things first ...

Perhaps we should also add a pinch of self-confidence to the discussion on our current business model; a refinancing model for journalistic content which still enables a return on sales well into double-digit figures, which other industries can only dream of, does indeed require a critical discussion on journalistic and publishing optimization in terms of strategic perspective, but it makes no practical sense if we set ourselves up cosily within a media concept anticipating the "demise of the print media." O. Spengler's "Demise of occident" meanwhile takes place nearly 100 years.

6 Other "business models" to finance editorial content

Incidentally, I cannot quite understand why financing through cooperatives or the government is supposed to "be good" for journalistic content. We are majority held (at 56 percent) by the *Publishing Group Georg von Holtzbrinck*, but the *Gesellschaft für staatsbürgerliche Bildung Saar mbH* (a non-profit organization) also effectively has a share of approx. 27 percent and the *Mitarbeiter-Beteiligungsgesellschaft*, which is owned by our employees, holds approx. 17 percent. While we are not a "cooperative undertaking," we are certainly one to put it reserved, with a not exactly homogenous shareholder structure. Against the background of my practical experience, I completely fail to see why the financing of content should be reorganized on the basis of public subsidies, cooperative financing or financing based on ownership rights when one is after all particularly concerned in the context of competition within the media industry about the quality and journalistic balance of news, analysis and (also regional) content. The public committees, as well as my own knowledge of the cooperative meetings of the *taz*, the Berlin based national newspaper owned by a cooperative, are in no manner likely to show any optimism in this regard. It may be that in a specific instance such arrangements could be relevant, but in my view they cannot be generalized. Even non-profit organizations must take the decision between distribution of dividends (may be for social purposes) and higher costs for editorial content.

How is it possible to finance thousands of journalists of the regional newspapers, which are needed to produce regular editorial content, by (public) funds (which have really nothing in mind but the quality and the balance of content)? Funds (with public or private shareholders) may be a solution for niche players, but not for regional newspapers (and other mass media) aiming at high range. Not to mention the fact that the advertising markets – contrary to their reputation – are far further removed from to stint editorial independence than the meetings of cooperative-society members and shareholders and in particular the model of government subsidies for which there is currently really no need.

7 The political framework for success

One last word on the financing of content: It cannot remain the case that, under the slogan "freedom on the high seas of the Internet," it is permissible to access free of charge newspaper content which has cost a great deal to produce, in order to be able to run one's own advertising business on search engines.

For this reason, the call by publishers for intellectual property rights must be heeded and the preconditions must thereby be created for negotiating on profit sharing with the search-machine operators, to be instigated either collectively or through government.

Neither can it be the case that public broadcasters backed with public finance have access to the digital newspaper market of the future without any legal restrictions.

Finally, we can finance journalistic content if we step forward to an "industrialization" of newspaper production. But we have to take into consideration high obstacles of our competition laws which exclude a lot of helpful solutions.

8 More focus, more scale …

When I began working in the steel industry in 1980, the demise of the sector was being heralded from all quarters. Gas pipes were in future to be made of plastic, cars only of carbon, and high-rise buildings were to be constructed without steel girders. The average production of raw steel has however been constantly fluctuating since this time at around 40 million tons. Only, the number of types of steel has increased to way above 5,000, and these are tailored to new uses, while production processes have been optimized to save resources and energy, and customer-oriented manufacturing processes are not just a "rhetorical exercise."

Now some people are coming to me and saying, "But Mr. Meinhold, we're not a steel factory … ". First: That may be the case, but the media industry can also learn something from the steel sector: The bait must appeal to the fish and not to the fisherman, meaning that if we have good content, we will also succeed in having a successful advertising business. However, our content is unfortunately no longer as "uncontroversial" as publishers would like to think and as they affirm to one another at conferences and when lobbying for intellectual property rights. Here, we will have to optimize if we wish to continue offering an attractive readership base and editorial environment to our advertising customers. And, secondly, there is learning from this area of business: Before we definitely archive a particular business model, we should after all exhaust all of its potential, particularly in view of its present still considerable success.

Part 3: **PR, Journalism, and Convergence**

Klaus Kocks
Exercising public influence

The interdependent system of public relations and journalism

Abstract: If there was supremacy of PR over journalism, PR would deny it. Based on this assumption, the following text describes the subtle and not always "measurable" ways highly professionalized PR is exercising public influence as well as the canting ways in which publishers, but also professional associations of journalists, deal with PR and contribute to the camouflage of PR. The conclusion is frightening. In an integrated system, the press becomes a function of business and political interests – though not necessarily the instrument of one specific party or company, as PR efforts may also counterbalance each other.

Keywords: public relations, corporate communication, fourth estate, spin doctoring, balance of power

---

1 On frogs and hummingbirds

Let us begin with that notorious modicum of truth in the information provided by the craftsmen about their craft. There is a German saying that goes: "When you have to drain a swamp you don't ask the frogs first." That is why journalists are notoriously bad witnesses when it comes to defining the state of the press. But we aren't asking the frogs – we're asking the swamp what it thinks of its inhabitants. That's an unusual approach, but it is conceivably also a pretty enlightening one, too.

Who controls whom? A complex question. An ornithologist once told me that the orange-red blooms of the strelitzia in the tropical rainforests control the hummingbirds, not the other way round. The plant controls the bird; when it wants to be fertilized the bloom changes color to attract the hummingbird, who succumbs to temptation, hovers over the arrow-like nectary with its wings whirring, pollinating as it drinks just enough nectar through its striking bill to replace the energy it has expended in flight. The ornithologist had a theory that the flowering plants created a bird that can hover and has an especially long bill for pollinating.

So who controls whom? An interdependent system with two mutually reinforcing functions. Which of the two has a public mission – the plants or the birds? The birds are definitely the ones doing the work. I have to admit it's all very confusing. But then you should never ask a cyberneticist who controls whom.

Back to my subject, though. Does PR reign supreme? And if so, is PR riddled with bad intentions? And how can PR shed some light on this dilemma? If there were supremacy, PR would deny it. And if PR were riddled with bad intentions do you really suppose that PR would actually admit it? Wouldn't the swamp where the frogs prosper, always pretend the frogs are the pinnacles of creation, and wouldn't the strelitzia in the rainforest pretend the hummingbirds set the tone?

2 PR, journalism, and the Garden of Eden

Public relations can be defined as communication management through persuasive communication based on an economically reinforced interest. That is not a discipline of self-criticism. And it is not an overly transparent craft. This is where the light is deliberately hidden under the proverbial bushel. Don't forget Theodore Roosevelt's motto: "Speak softly, but carry a big stick!" This is too stiff a challenge for me – how am I supposed to determine the quintessence of PR?

The second stiff challenge is: What about the journalistic world that surrounds us? What media are we talking about? Is it the ordered world once mirrored by page one of the *Frankfurter Allgemeine Zeitung*? With a distinction between news and opinion, report and commentary, news desk and advertising? Was that kind of compartmentalization not swept away with the advent of the feature, or at the very latest, the Internet? It is a painful but true fact that – in the eyes of the younger generation – journalism is no longer a profession. We have been taught to believe that the value of Wikipedia is rooted in the principle that the writers are not experts; expertise from the pen (or rather, keyboard) of non-experts, that is modern-day *Leitkultur*.

And finally, the difference between journalism and PR. A journalist in the PR business is like a cheap Polish tiller on a construction site. But that is no longer an individual misdemeanor, it has become collective business practice. Today, hardly any publishing house or newsroom doesn't offer PR services and, behind the scenes, corporate publishing and editing dovetail in many different ways.

If you listen to the guardians of the fourth estate it is as if time has gone haywire. They talk as if original sin had not yet been committed, as if Adam could still be persuaded to abstain from taking a bite out of Eve's apple. But – as they say in Parliament – out there in the country it isn't the Garden of Eden – out there in the country, Sodom and Gomorrha rule.

3 The increasing influence of PR

I am digressing into the realms of polemics, a superfluous exercise in this context. In any case, the theory of rhetoric teaches us that a good text should begin with a *captatio benevolentiae*, a technique that seeks to capture the attention of the reader. So I should have begun by flattering. Although I have baulked at doing that all my life, I am persuaded it is appropriate to make an exception here.

If the press is credited with a public duty, or – to phrase it another way – a general political role aka the fourth estate, it is becoming increasingly dependent on PR, but cannot per se be controlled by a single PR. Does PR control the press? Yes and no. We are talking about a relatively autonomous, yet multi-causal system whose overdetermination is increasingly attributable to PR.

The influence of public relations on the free press, if we take that to mean independent newsrooms, is growing. In fact, this influence has acquired a dimension where it would appear appropriate to refer to a systemic breakdown. That is backed by empirical evidence based on various methods and thus of varying quality. You can compare the number of press releases with the number of articles about an event or issue published in the press. Or conduct a headcount of the number of "press spokespersons" versus the number of "journalists." Or mold the sensitivities of journalists as determined by surveys into a normative canon. Or count the infringements of codes of conducts in press, publicity or PR watchdogs. All of that, however, although it is academic practice, is pretty superficial.

The truly fundamental issue in an economic context is the industrial rationalization of newsrooms pursued by publishers and the willful externalization of editorial costs in the direction of so-called content providers where it is common knowledge that their free deals are financed by a third source. In other words, there is a third interest in the equation. Newsrooms cost money, PR comes for free, and that's the publisher's gospel. Freedom of the press has a clear enemy, namely publishers. And the press council is, if I may exaggerate slightly, the publishers' PR agency tasked with muddying the systemic facts with accidental criticism.

4 Motor, business, and political journalism

Travel and motor journalism is commonly discredited as being PR-controlled, but that is not merely cheap hypocrisy, it is arrogant to boot, particularly on the part of the esteemed colleagues of the business and features desks and, above all, the bigwigs of the political news desks. Generally speaking, motoring journalists have actually driven the car they are reviewing. In pleasant circumstances, admittedly, but nevertheless "for real." No one could claim the same for the business or financial press, or for the political punters, where all too often semi-qualified amateurs comment on products they don't really understand. There are myriad examples from the last financial crisis. Motoring journalists at least have a driving license. However, the dearth of driving skills among those taking part in the stock exchange rallies is merely a peripheral problem. There is worse to come.

One of the key problems for the business media is that all news agencies are in the hands of the people doing the business. Stamocap rules the day. For me, Bloomberg is a paradigmatic example: broker first, news agency second, finally mayor. Business, media, and political power all rolled into one. I could extemporize and extend the Bloomberg paradigm to Berlusconi. Bunga-bunga all the way so that we don't ask the uncomfortable and critical questions. And I could expand on that to embrace the self-defense of the Murdoch Empire.

My objection though, is more fundamental than a particular shade of PR. That's only cosmetic. My objection concerns the systemic integration of the business, politics and press functions in the media system. Let me use a metaphor from the world of sports betting, because the stock exchange is in essence little more than a betting platform. How come the players who are playing the game can bet on the outcome of that game? Do they bet on how they will play? Or do they play like they bet? That question is of fundamental relevance for capital markets which decide on the fate of entire nations.

In an integrated system the press becomes a function of business and politics. Not always the politics of one specific party. But ever more frequently a function. If PR used to operate in a sellers' market, PR today is increasingly operating in a buyers' market. That is the fundamental economic and publicistic shift. It is the homeland of one particular polemic expression: A PR manager can never tell enough lies to satisfy journalistic demand.

5 PR is looking for a healthy host

Let me phrase it more carefully: Work-shy yet self-opinionated journalism has always claimed as its natural right that stakeholders should communicate their interests in the form of suitably packaged news bites ready for publication. This began by first claiming the duty of authorities to inform, and was then transferred to companies and extended to cover a right to original sound bites and TV interviews. Otherwise, the virtuous indignation of the yellow press knows no bounds. As the influence of capital markets grows, the claim also extends to the shareholders' right to know what's going on. In a political context this is a culture of participation that sits cowering on the shoulders of representative democracy. Consequently, a new profession is born – that of the pre-journalistic information "tweaker," popularly known as a press spokesperson or a PR manager.

That is a special type of economic phenomenon which calls for further reflection: Because this service is not paid for by the user, but by the provider, that implies its intrinsic economic value must consist of an additional benefit – disguised to a greater or lesser degree – for the provider himself. And that is why the polemic metaphor that PR is a parasite of the free press is apposite; a healthy host is most welcome. The readers, society, the community at large pick up the tab.

An issue of relevance for the general state of the system is whether the influence of PR is a core function or just an occasional glitch in an otherwise intact system (underdetermination), or whether the functionalization of journalism has reached such a stage that the surface scaffolding no longer corresponds to the underlying structures (overdetermination). Looks like press, but in fact it's PR. The reliability of information is then merely the consequence of various PR effects that in an ideal world balance each other out. Or culminate in chaos that turns a blind eye to dictatorial harmonization. The checks and balances thus shift to PR, which in essence becomes the regulator. In the publicist system, the watchdog function switches from the relatively autonomous newsrooms to the diversity of the PR offering, where many have a great opportunity to slash specific costs; in the Web the cost curve almost hits zero. "Have your say": That is the new imperative for everyman.

My argument follows second-order cybernetics, but I'll spare the scientific jargon. Journalism itself is discussing this structural change using the imagery of decadence, a decay in moral standards that it is our ethical duty to resist – to quote the title of a lachrymose book on self-reassurance penned by a journalist, where PR is painted as Beelzebub, to be resisted at all costs. Tut-tut – the editors are conjuring up miniature devils of temptation and gloating over their independence from these imaginary figures. Not good enough.

What really needs defending is the professional privileges of journalists, including acceptable minimum working conditions and union affiliation. The *Deutscher Journalisten-Verband* (the German Federation of Journalists) is nowhere near to meeting its obligations. The federation is infiltrated and controlled by PR. A bad joke by pretentious apparatchiks. The decline of publicists is evidenced by the fact that there is no serious discussion of this downfall in Germany. The only voices that carry any weight are to be heard in *Netzwerk Recherche*, a German association of investigative journalists, which came into being as a grand gesture in defiance of journalistic day-to-day reality.

6 Identity, interest, ideology, and intention

But I don't want to evade the issue. I have been asked to comment on whether PR is a danger to the free press in that it controls the media. I have four answers to that question:

1) PR and journalism are the opposite sides to the communications coin. Publishers must set up their newsrooms so that they can use PR, but not succumb to it because they have been deprived of adequate working conditions. From a journalist's perspective it is good to know about a bias of interest, but then, as such, any bias of interest is problematic and therefore always requires examination. Every source is suspect. Any good historian knows that. And in the interests of the opinion-forming process for the citizens of this country, making PR a common cause is out of the question, no matter how noble the cause may be.

2) PR does not present a regulatory problem for as long as identity, interest, ideology, and intention are clear. That is my theory of the four "I"s. But it is precisely because PR seeks to avoid this transparency systematically, not merely accidentally, that PR also poses an accidental problem. Every source is suspect. There is no such thing as making a common cause out of PR, or a transparent one, either.

3) The question of whether there is still a level playing field for press spokespersons and journalists or whether the system is collapsing is of such importance that it has to be answered specifically and on a case-by-case basis. We are all aware of cases where system functionalisation brings far-reaching changes in underlying structures so that the true workings of a system can no longer be inferred from its look and feel. Every claim to truth is suspect. Truth is not a common cause.

4) The political animals in the PR world and those academics who compliantly pocket the proffered contributions are vehemently opposed to my viewpoint. The profession's ethical codes of conduct are making every effort to mold PR into an instrument that fosters the truth and is an original expression of democratic culture. A watchdog is to sanction any infringement. But that isn't self-regulation. It is the attempt to muzzle the critics from their own ranks. I myself have observed the work of the watchdog chiefly in the efforts to censor the present comments. Because the real agenda is PR for PR. Because this is all about credibility. The fifth estate is puffing itself up against the fourth. If what the fifth estate says about normative guidelines for PR were appropriate, newsrooms could pack up and go home now. If what the fifth estate says were true, we would in future be asking Doctor Marlboro whether smoking is

healthy. And Angela Merkel whether the government is up to it. Maybe then we should name the newspapers "pravda," or "truth."

Those are my four theses. As you can see, I'm putting my own house in order, too.

I call on my profession to respect the four "I"s of identity, interest, ideology, and intention. And I say to journalists: No matter where, no matter when, those – and none other – are the rules of the game. Not by accident or coincidence, but systemically.

7 PR has become a buyer's market

But let us return to the teetering system: All newsrooms large and small are jostling their way into this game by offering PR as an in-house service. And blatantly selling editorial expertise as consulting under the guise of an editorial cover-up.

Whether individual cases surrounding prominent PR managers are singled out for the headlines or not, whether these cases muddy the distinction between professional consulting and organized crime does not impinge on my position. I gave everyone plenty of warning. It is a problem for those editors-in-chief who scurry around the spin doctors. The chiefs are said to pay court to these shadowy figures because – and here comes the argument already mentioned – PR is no longer a sellers' market but a buyers' one. To put this into more precise economic terms, spin doctoring has assumed the role of a coveted supplier whose raw materials are urgently needed to manufacture the required product because precisely this product is in demand on an intensely competitive fake market. Disinformation is avoided where possible, but it's better to run a bad story than leave it to the competition. Journalists must put that house in order.

8 The press as a form of rhetoric

Finally, I want to offer a positive description of what the press is, namely a form of public address that must meet special requirements. In terms of systematics, the press, like PR, like advertising, is a form of rhetoric. The press isn't a scientific institution so it is not subject to the criteria of probity and veracity. The press is rhetoric. Ever since the Ancient Greeks, rhetoric has been understood to mean forms of public address designed to persuade or influence. The press as we have come to understand it in recent centuries fulfills a special function, represents a particular form of rhetoric.

However, in the modern business world there was demand for "reliable news" in addition to the rumors and stories, because everything else could lead to ruin. And so the modern press was born – "correct, reliable, but terse." A special species that prevents the business community from falling for every crazy idea or every rumor. The press rises as the phoenix of common interest from the charred ashes of speculation. It is supposed to make up the threesome with the businessmen and the storytellers. It can perform that task if everyone else does the same in a community spirit. And if everyone else is willing to pay adequately for this service.

The citoyens must be allowed to supersede the bourgeoisie in the press, even if we are talking about one and the same group. And in that sense, the press is not the fourth estate, a function of the common good (volonté générale), but merely a function of the common interest (volonté de tous), and to infringe on that can be the essence of good business. Citoyens and bourgeoisie are opponents, not partners. To seek to unite PR and press in an intereffication model is tantamount to a counter-enlightenment maneuver.

Democracy isn't a bed of roses.

What am I propagating? Kant makes a distinction between the public and private uses of reason. Kant uses the term private to mean the corporate sector. The journalist sells his or her articles or newspapers, the press spokesperson sells his or her line. Overlying that is the public debate on what we as enlightened people believe to be reasonable. What we consider to be right and good in the way of general laws. What our moral principles demand of us.

Barbara Baerns
A changing interplay?

Public relations and journalism in a converging media world

Abstract: In an attempt to review the current state of research concerning the interplay of journalism and public relations as well as the converging media world, this article is complementary to Klaus Kocks' contribution to the subject. Kocks advances the notion of latency being a vital characteristic of public relations in general. In this article however, latency is considered to be the result of individual decision-making and strategic intervention in the areas of journalism and public relations. The main objective for research in this field is to expose latent relations and influences, and thereby create more transparent media coverage and a more transparent media system. So far, there has been little research that has examined the interplay of public relations and online journalism. Even Kocks does not consider the possible changes to journalism and public relations as a result of the new technology.

Keywords: public relations, journalism, converging media, diversity of media, transparency in journalism

1 Introduction

Until recently, when discussing the difference between journalism and public relations (PR) and how they interact, I focused on findings concerning offline media and news agencies (Baerns 2007; Baerns 2009). Looking back to older data collections is helpful to pinpoint the changes in these areas. To provide a better understanding of the nature of these developments, this article focuses on three aspects, namely the self-concept of the communicators, the content, and verification possibilities and criteria. The findings of this review should make it possible to identify the need for further research and to develop action plans within the field.

2 The self-concept

Surveys carried out in the field of journalism show that the latter is stable, as far as journalists' understanding of their roles and the practical consequences of self-perception are considered. Most journalists reject the idea of establishing close ties with public relations. A comparison of the recent survey *Journalism in Germany*, which presented data from 2005, as well as an earlier survey with data collected in 1993, confirms this finding (Löffelholz 1997; Scholl/Weischenberg 1998; Weischenberg/Malik/Scholl 2006). Nevertheless, the authors mention difficulties in differentiating between journalism and public relations work (Weischenberg/Malik/Scholl 2006: 346 f.). Journalists are still cautious about their PR relations with more than half seeing PR as an alternative form of marketing. This is demonstrated by the results of a survey of specialized journalists (100 per sector) covering the last ten years (Com.X/prmagazin 2010).

However, the vast majority of journalists (86 percent) agreed with the notion that professional PR helps journalists work efficiently. Nearly all journalists covering business and economics (90 percent) view themselves as "PR clients, who insist on swift and quality service" (Com.X/prmagazin 2010). The project, *Actors, structures and achievements in and of today's Internet public*, run by Christoph Neuberger and funded by the German Research Foundation (DFG), involves interviews with editors-in-chief. Unfortunately, the study did not explore the links between journalism and PR, focusing instead on how weblogs may influence journalistic content (Neuberger/Nuernbergk/Rischke 2009). Machill, Beiler and Zenker (2008) explored how editors use the Internet for supplementary research, and to which extent they rely on the Internet to gather (not verify) additional information. This research has shown that editors primarily use the search engine Google and online journalistic content, rather than consult primary sources such as the websites of companies or administrations. Because Google itself draws on the same sources, journalism tends to become self-referential, as Machill, Beiler and Zenker argue (2008: 291 ff.). The same authors published a how-to book offering advice on the use of search engines (Machill/Beiler/Gerstner 2009).

Strangely enough, the description of their profession by the German Journalists Federation (Deutscher Journalisten-Verband, DJV) still extends no further than the actual practical work performed by journalists. In contrast, the German Public Relations Association (Deutsche Public Relations Gesellschaft, DPRG) continues to define PR as the process of managing (commissioned) communication. However, working with the press as well as with other forms of media is the top priority in the field of public relations – in terms of both the time and commitment invested. This has been proven time and time again, since the first survey of German PR experts in 1973 (DIPR 1973). The latest survey (Szyszka/Schütte/

Urbahn 2009) showed that having an Internet presence clearly comes second after working with the media (84 percent as compared to 93 percent). This is despite the fact that e-mailed press releases have long since replaced hard copy press releases. Online PR (65 percent) ranks tenth out of 22 predefined items (Szyszka/ Schütte/Urbahn 2009: 123). Journalists are considered the most important target group, even though PR could easily use the Internet to circumvent traditional media and communicate directly with their respective stakeholders and publics.

This finding has also been confirmed by the results of surveys of communications and press officers from the Federal Association of German Press Spokespeople (Bundesverband Deutscher Pressesprecher), which have been conducted every two years since 2005 (Bentele/Grosskurth/Seidenglanz 2005 ff.). On the other hand, in 2005 only 35 percent, and in 2007 only 36 percent of those surveyed said they expected journalists to have a high or very high degree of confidence and trust in the PR sector. For the 2009 survey, some of the questions were modified. The press officers were asked to rate to what extent journalists trust their individual PR work, which led to strikingly different results. The latest survey, which was originally due to be released in 2011 according to the publisher, is currently being redesigned – also with regard to online communication. There is also a considerable amount of monitoring of social networks (Selbach 2011). In 2012, the PR-Trendmonitor survey predicted a stronger presence of PR departments and agencies on social media sites and more press releases via Web 2.0.

It is worthwhile to reiterate that from the beginning of this research right up to the present day, survey methods have produced results that favor differentiation according to perceptions, roles, and functions of journalists and PR professionals. However, concrete activities may differ.

3 The content

In the years following the author's first publications, research based on consistent designs[1] proved that PR had a considerable influence on which topics

---

1 A study of latent relations requires a reasonable framework for the purpose of delineation. Applying the relevant normative frames, the distinction between journalism and public relations was based on the assumption that there is, in fact, a functional difference between journalism and PR, both being conceptualized as "information systems which are equivalent in terms of syntax, but not semantics" (Baerns 1991: 16). Both information systems were conceived of as competing forces; striving for primacy in defining media content and influencing public discourse ("Deutungshoheit").

make it into the media and when (*The power of public relations*) (Baerns 1991). It was shown that news agencies were the most likely media to reveal their PR sources, with television stations at the other end of the spectrum.[2] On the basis of these findings, the study followed up on the notion that media techniques and media dramaturgy, i.e. forms of mediatization have a considerable impact on the content of media coverage (*The power of the media)* (Baerns 1987). The case studies conducted on the matter corroborate the notion that media structures and journalistic narration routines (such as the climax first principle; using image sequences to illustrate predefined texts, or paraphrasing) affect the meaning of the content, irrespective of the editorial line or the views of the journalist. It was shown that distortions in the media system already begin in the newsroom of a news agency. Factual errors made by a news agency also often affect headlines and leads. Instead of being corrected, such errors accumulate when the information is passed on through the media system.

Klaus Kocks' cybernetic approach (see Kocks in this book) also focuses on the interaction (i.e. the mutual control of journalism and public relations via feedback loops) within a system that ought to be in balance – but is in danger of collapse. Restricted as it is to the macro-level, this approach can neither help us gain applicable knowledge on individual selection and transformation processes, nor does it contribute towards a better understanding of the impact of the disseminated (dis)information.

Though public relations professionals maintain that the interplay between PR and journalism ensures an optimal supply of reliable information (Rolke 1999), PR and journalism have not succeeded in achieving this goal. Intentionally or not, the result of this interplay is an overall media performance which does not live up to the expectations of our democracy. The PR sector probably has adapted to this situation. For years, several PR experts have continued to champion the argument that mass media creates the necessary noise. If you actually want to communicate, use different tools and channels. It remains to be seen how, in the long run, the dialogical capacity of the Internet will be part of PR strategies. There is however no empirical evidence that communication in the real sense of the word has been the main PR-objective (Baerns 2008).

A first trawl of the Internet suggests there is a need for empirical studies on both "straightforward" public relations and latent public relations through

---

2 The latest study carried out on Germany's news agency dpa (participant observation, content analysis, a 6-week long survey in 2004/2005) came to the conclusion that one in four publications are falsely declared to be based on original research, with the real sources being disguised (Höhn 2005).

"pseudo-journalism" (Neuberger 2002). However this article may meet objections which stand and fall on whether the reader accepts that a new communication environment must not necessarily supersede traditional interpretative models. Thus Nothhaft recently questioned, "whether Barbara Baerns' determination thesis still deserves to be considered (at all) given the drastic changes in the media landscape" (Nothhaft 2012: 254). This notion corresponds with the assertion that the new environment has fundamentally changed the investigative and gatekeeping functions of journalism – or has even rendered them unnecessary. A more convincing argument is, due to the sheer vastness of the converging media system, that the lines between public relations and journalism have become so blurred that it is no longer possible for research to distinguish between them. Recent studies distinguish between websites offering online journalism and those providing non-journalistic content (e.g. Beck/Dogruel/Reineck 2010). Their focus is on which topics, not whose topics make it onto the Web (e.g. Neuberger/ Lobigs 2010; Zeller/Wolling 2010). They focus on users, and claim transparency of content rather than transparency via presentation as a criterion of quality (e.g. Siegert/Trepte/Baumann/Hautzinger 2005).

This, however, is not the main objective regarding the differentiation between journalism and public relations. Headed by Miriam Meckel, a research group at the Institute for Media and Communication Management at the University of St. Gallen was the first to use a research design combining network- and content analysis for their exploratory case study on the use of the communication platform Twitter as a source of information. The great advantage of this research design is that it also provides an opportunity to systematically observe the diffusion and transformation of PR information on the Internet (Plotkowiak et al. 2012).

Adequate presentation is another subject that merits more research. Ever since the publication of the first findings pointing to a latent interplay between PR work and journalism, researchers have demanded that journalists provide information on the genesis of their contributions. In practice, this demand was often rejected as incompatible with the binding narration routines of journalism. Questions concerning how and to what extent traditional media and online portals harness the opportunities provided by new formats, create a new issue for observation and discussion.

4 Verification possibilities and criteria

The notion that a large number of different media outlets alone will guarantee the diversity of media content has gone along with the structural development of our media system. For decades, the rulings of the German Federal Constitutional Court gave priority to structural pluralism. The federal government, in its Media Reports from 1970 up to the latest edition from 2008, has always acted under the assumption that there is a link between media pluralism and information pluralism. This link is supposedly established by the recipient, who draws on information from different media outlets and compares it in order to form an independent opinion on the matter at hand. This concept was first explained in the late 1960s (Michel Kommission 1967), and has been adopted within the scientific community. Elisabeth Noelle-Neumann, for instance, argues in a similar vein. For her, the main reason why Nazi propaganda proved to be effective was the limited choice of media content (Noelle-Neumann 1971: 344). For Franz Ronneberger, "having a real possibility to choose" is one of the central sociological and political aspects of the concept of "freedom," which he explores in his classification of communication policies (Ronneberger 1978: 191 f.): "It seems that journalistic pluralism is not an end in itself, but a way of finding the truth" (Ronneberger 1978: 229). Ever since the publication of the "Four Theories of the Press" (Siebert/Peterson/ Schramm 1956), pluralism has been regarded as a central feature of democratic media systems. This approach has also been adopted by researchers working in comparative media studies and transformation studies (e.g. Jarolimek 2009). Pluralism remains a referential criterion among others within the Media for Democracy Monitor (MDM), which was initially implemented in 2006 (Trappel/Nieminen/Nord 2011: 29 ff.).

The validity of the model was first tentatively questioned thirty years ago, during the debates on the concentration of media ownership (Klaue/Knoche/ Zerdick 1980: 80 f.). This inspired the author's own research on the interplay of PR work and journalism in the media system of North Rhine-Westphalia, and gave rise to the idea of empirically verifying evidence. Using this research design did not only offer an approach to prove the pluralist model to be invalid, it also made it possible to expose the inner workings of a system that pretends to be pluralistic, but in fact relies on copying and distributing information derived from the same sources – in this case political PR (Baerns 1983). As the origins of the material at hand remain opaque, the failed attempt to "structurally secure the desire for truth" is inconsequential. Put more precisely: It is the lack of transparency regarding the sources, it is the functionality of dysfunction that keeps the overall concept alive.

So far, there have not been any studies into the workings of the system. The comparative program analysis carried out under the conditions governing the

German Dual Broadcasting System[3] did not look at the sources of broadcast material (for the relevant empirical studies: Zeller/Wolling 2010: 144, footnotes 10 ff.).
However, the results of the recent study on *The role of the Internet for ensuring
pluralism*, commissioned in 2009 by the German Commission on Concentration
in the Media (Kommission zur Ermittlung der Konzentration im Medienbereich,
KEK) are consistent with the author's previous findings:

> Contrary to a widely spread supposition that a large number of information providers auto-
> matically leads to a high level of content diversity, this is not the case. (...) If we only look
> at online journalism, we see that the market is overwhelmingly dominated by traditional
> mass media, which continue to use their online presence to recycle journalistic content
> from the parent medium on a large scale. (...) In addition to this, we find that there is strong
> co-orientation among information providers, meaning that topics covered by one supplier
> are picked up on by others, which in turn decreases content diversity. Also, some informa-
> tion providers work together, though little is known about the extent of these co-operations.
> (Neuberger/Lobigs 2010: 166)

Neuberger's summary is also consistent with Beck's findings, according to which
online business coverage does not create any journalistic added value (Beck/
Dogruel/Reineck 2010: 248 f.).

In light of these findings, it is hardly surprising that there have been recent
calls for new transparency in journalism itself. Meier and Reimer (2011) put
forward the following three-tired transparency model: At the product level,
journalists have to reveal their sources and the latter's interests. At the process
level, editors should explain and justify their choice of topics, news placement
decisions and their editorial stance on topics. At the third level, journalists and
editors interact with users, especially and most conveniently, over the Internet.

5 Conclusion

In conclusion, it is worthwhile to point out that not only the future of journalism, but also that of PR remains unpredictable. Both co-existing professions
find themselves in a challenging position. The new technology provides public
relations professionals with the proper means to close the gap between what is
preached and what is practiced – communication. Even though it did not seem
altogether reasonable in the past, the traditional focus on journalism by prac-

---

3 The current broadcasting system in Germany allows for public and private radio and television
stations to co-exist and is therefore referred to as the Dual Broadcasting System.

titioners appears to be inevitable in order to enter a self-referential journalistic system.

Looking at the other side of the coin, journalism, it becomes clear that the field is burdened with a legacy of shortcomings,[4] preventing professionals from tapping into the innovative potential offered by online communication. The process of differentiation remains incomplete. To efficiently counteract "counter-enlightenment maneuvers" (see Kocks in this book), journalists' independence and credibility will not have to be safeguarded but in fact needs to be developed in the first place.

References

Baerns, Barbara. Vielfalt und Vervielfältigung. Befunde aus der Region – eine Herausforderung für die Praxis. *Media Perspektiven* 3 (1983): 207–215.

Baerns, Barbara. Macht der Öffentlichkeitsarbeit und Macht der Medien. In *Politikvermittlung. Beiträge zur politischen Kommunikationskultur. Schriftenreihe der Bundeszentrale für politische Bildung (238)*, Sarcinelli, Ulrich (ed.), 147–160. Bonn: Verlag Bonn Aktuell, 1987.

Baerns, Barbara. *Öffentlichkeitsarbeit oder Journalismus? Zum Einfluss im Mediensystem.* Revised ed. (1st ed. 1985). Köln: Verlag Wissenschaft und Politik, 1991.

Baerns, Barbara. The "Determination Thesis": How independent is journalism of public relations? In *A Complicated, Antagonistic & Symbiotic Affair. Journalism, Public Relations and their Struggle for Public Attention*, Merkel, Bernd/Russ-Mohl, Stephan/Zavaritt, Giovanni (eds.), 43–57. Lugano: Università della Svizzera Italiana, Giampiero Casagrande editore, 2007.

Baerns, Barbara. Public relations is what public relations does. Conclusions from a long-term project on professional public relations. In *Public Relations Metrics. Research and Evaluation*, Van Ruler, Betteke/Tkalac Vercic, Ana/Vercic, Dejan (eds.), 154–169. New York, London: Routledge, 2008.

Baerns, Barbara. Öffentlichkeitsarbeit und Erkenntnisinteressen der Publizistik- und Kommuni-kationswissenschaft. In *Theorien der Public Relations. Grundlagen und Perspektiven der PR-Forschung*, Röttger, Ulrike (ed.), 285–297. 2nd revised ed. Wiesbaden: VS Verlag für Sozialwissenschaften, 2009.

---

4 Notably the following: (1) The prerogative of information gathering granted to journalists by the Media Laws was rarely used. Meanwhile, the so-called Freedom of Information Act has extended this right to all citizens. (2) Even the Journalists' Association Netzwerk Recherche e.V. does not distinguish between the demand for a revelation of journalistic sources on the one hand and the protection of those who provide journalistic information on the other. (3) There is currently no debate on whether the journalist's profession should be protected in Germany. Germany has no specific Journalism Act. (4) Journalists are reluctant to accept research findings. Bridging the gap between researchers and practicing journalists remains a challenge.

Beck, Klaus/Dogruel, Leyla/Reineck, Dennis. Wirtschaft Online: Zweitverwertung oder publizistischer Mehrwert? Eine Analyse aus Kommunikatorsicht. *Publizistik* 55 (2010): 231–251.

Bentele, Günter/Grosskurth, Lars/Seidenglanz, René. *Profession Pressesprecher: Vermessung eines Berufsstandes.* Berlin: Helios Media, 2005, 2007, 2009.

Com.X/prmagazin. Entspannte Verhältnisse? *Prmagazin* 41, no. 12 (2010): 34–39.

Deutscher Bundestag. Zwischenbericht der Bundesregierung über die Lage von Presse und Rundfunk in der Bundesrepublik Deutschland. Drucksache 6/692, 1970.

Deutscher Bundestag. Unterrichtung durch den Beauftragten der Bundesregierung für Kultur und Medien. Medien- und Kommunikationsbericht der Bundesregierung. Drucksache 16/11570, 2008.

DIPR (Deutsches Institut für Public Relations). Primärerhebung. Berufsbild Public Relations in der BRD. Köln: Unpublished Manuscript, 1973.

Höhn, Tobias D. Schnittstelle Nachrichtenagenturen und Public Relations – eine Untersuchung am Beispiel der Deutschen Presse-Agentur (dpa). Unpublished Master's Thesis University of Leipzig, 2005.

Jarolimek, Stefan. *Die Transformation von Öffentlichkeit und Journalismus. Modellentwurf und das Fallbeispiel Belarus.* Wiesbaden: VS Verlag für Sozialwissenschaften, 2009.

KEK (Kommission zur Ermittlung der Konzentration im Medienbereich). Press Release August 2009. Ausschreibung der KEK für ein Gutachten zum Thema „Die Bedeutung des Internets im Rahmen der Vielfaltssicherung". Retrieved on 19th August 2009, from www.kek-online.de (2009).

Klaue, Siegfried/Knoche, Manfred/Zerdick, Axel (eds.). *Probleme der Pressekonzentrationsforschung. Ein Experten-Colloquium an der Freien Universität Berlin. Materialien zur interdisziplinären Medienforschung 12.* Baden-Baden: Nomos, 1980.

Löffelholz, Martin. Dimensionen struktureller Kopplung von Öffentlichkeitsarbeit und Journalismus. Überlegungen zur Theorie selbstreferentieller Systeme und Ergebnisse einer repräsentativen Studie. In *Aktuelle Entstehung von Öffentlichkeit. Akteure – Strukturen – Veränderungen,* Bentele, Günter/Haller, Michael (eds.), 187–208. Konstanz: UVK, 1997.

Machill, Marcel/Beiler, Markus/Zenker, Martin unter Mitarbeit von Gerstner, Johannes R. *Journalistische Recherche im Internet. Bestandsaufnahme journalistischer Arbeitsweisen in Zeitungen, Hörfunk, Fernsehen und Online. Schriftenreihe Medienforschung der Landesanstalt für Medien Nordrhein-Westfalen 60.* Berlin: Vistas, 2008.

Machill, Marcel/Beiler, Markus/Gerstner, Johannes R. *Online-Recherchestrategien für Journalistinnen und Journalisten. Workshopmaterialien für die Aus- und Weiterbildung.* Edited by Landesanstalt für Medien Nordrhein-Westfalen. Wuppertal: Boerje Halm, 2009.

Meier, Klaus/Reimer, Julius. Transparenz im Journalismus. Instrumente, Konfliktpotentiale, Wirkung. *Publizistik* 56 (2011): 133–155.

Michel Kommission. Bericht der Kommission zur Untersuchung der Wettbewerbsgleichheit von Presse, Funk/Fernsehen und Film. Bundestags-Drucksache 5/2120, 28th September 1967.

Neuberger, Christoph. Alles Content, oder was? Vom Unsichtbarwerden des Journalismus im Internet. In *Innovationen im Journalismus: Forschung für die Praxis,* Hohlfeld, Ralf/Meier, Klaus/Neuberger, Christoph (eds.), 25–69. Münster: Lit, 2002.

Neuberger, Christoph/Nuernbergk, Christian/Rischke, Melani. Journalismus im Internet: Zwischen Profession, Partizipation und Technik. *Media Perspektiven* 47 (2009): 174–188.

Neuberger, Christoph/Lobigs, Frank. *Die Bedeutung des Internets im Rahmen der Vielfaltssicherung.* Gutachten im Auftrag der Kommission zur Ermittlung der Konzentration im Medienbereich (KEK). Berlin: Vistas, 2010.

Noelle-Neumann, Elisabeth. Wirkung der Massenmedien. In *Publizistik*, Noelle-Neumann, Elisabeth/Schulz, Winfried (eds.), 316–350. Frankfurt am Main: S. Fischer, 1971.

Nothhaft, Howard. Rezension zu Ulrike Röttger, Joachim Preusse und Jana Schmitt: Grundlagen der Public Relations. Eine kommunikationswissenschaftliche Einführung. *Publizistik* 57 (2012): 253–254.

Plotkowiak, Thomas/Stanoevska-Slabeva, Katarina/Ebermann, Jana/Meckel, Miriam/Fleck, Matthes. Netzwerk-Journalismus. Zur veränderten Vermittlerrolle von Journalisten am Beispiel einer Case Study zu Twitter und den Unruhen in Iran. *Medien & Kommunikationswissenschaft* 60 (2012): 102–124.

PR-Trendmonitor. PR-Trend 2012. *PR Report* 12 (2011): 11.

Rolke, Lothar. Journalisten und PR-Manager – eine antagonistische Partnerschaft mit offener Zukunft. In *Wie die Medien Wirklichkeit steuern und selbst gesteuert werden,* Rolke, Lothar/Wolff, Volker (eds.), 223–247. Wiesbaden: Westdeutscher Verlag, 1999.

Ronneberger, Franz. *Kommunikationspolitik I. Institutionen, Prozesse, Ziele*. Kommunikationswissenschaftliche Bibliothek 6. Mainz: v. Hase & Koehler, 1978.

Scholl, Armin/Weischenberg, Siegfried. *Journalismus in der Gesellschaft. Theorie, Methodologie und Empirie*. Wiesbaden: Westdeutscher Verlag, 1998.

Selbach, David. Lauschangriff. *Prmagazin* 42, no. 10 (2011): 60–65.

Siebert, Fred S./Peterson, Theodore/Schramm, Wilbur. *Four Theories of the Press. The Authoritarian, Libertarian, Social Responsibility and Soviet Communist Concepts of What the Press Should Be and Do*. Urbana, Illinois: University of Illinois Press, 1956.

Siegert, Gabriele/Trepte, Sabine/Baumann, Eva/Hautzinger, Nina. Qualität gesundheitsbezogener Online-Angebote aus Sicht von Usern und Experten. *Medien & Kommunikationswissenschaft* 53 (2005): 486–506.

Szyszka, Peter/Schütte, Dagmar/Urbahn, Katharina. *Public Relations in Deutschland. Eine empirische Studie zum Berufsfeld Öffentlichkeitsarbeit*. Konstanz: UVK, 2009.

Trappel, Josef/Nieminen, Hannu/Nord, Lars (eds.). *The Media For Democracy Monitor. A Cross National Study of Leading News Media*. University of Gothenburg: Nordicom, 2011.

Weischenberg, Siegfried/Malik, Maja/Scholl, Armin. Journalismus in Deutschland 2005. Zentrale Befunde der aktuellen Repräsentativbefragung deutscher Journalisten. *Media Perspektiven* 44 (2006): 346–361.

Zeller, Frauke/Wolling, Jens. Struktur und Qualitätsanalyse publizistischer Onlineangebote. Überlegungen zur Konzeption der Online-Inhaltsanalyse. *Media Perspektiven* 48 (2010): 143–153.

Marcello Foa

Journalists know little about spin doctors: This is the problem!

How the news agenda is infiltrated by hidden propaganda

Abstract: Journalists like to think of themselves as watchdogs of the public interest or as the fourth estate in a democracy. However, politicians have learned how to undermine the power of journalists by using the (alleged) supremacy of news media for their own purposes. They hire spin doctors who understand the inner workings of news organizations very well and who use sophisticated techniques to manipulate the news agenda in their client's interest. They twist facts and turn news into propaganda. Although spin has mainly worked in the Anglo-Saxon world, in an age of globalization and communication politicians in Continental Europe are increasingly using spin doctors' techniques as well. Therefore, journalists need to counter the spin in order to preserve their credibility.

Keywords: spin doctoring, media manipulation, transparency, media and democracy, public relations

---

1 A complex reality

Journalists like to portray themselves as watchdogs of the democratic system, serving the public interest. According to a widespread dogma in our age – the information age – politicians are dominated by the media. But are we truly in the so-called "mediocracy" (a democracy where media outlets are much more influential and effective than voters)?

My answer is no. We live in a more complex reality where apparently media have a lot of power but where politicians also have learned how to use the (alleged) supremacy of news media for their own purposes. How do they do that? It is done by employing communication experts, an unsurprising and inevitable trend which results in serious implications for public life. However a fundamental distinction needs to be made between consultants who act correctly by applying licit techniques and those who aim not to inform but to manipulate the media and the public opinion. The latter can be described as spin doctors (Foa 2006).

Truthful communication experts know and respect the boundary between institutional communication (public affairs) and political communication. Institutional communication should be neutral, reliable, and truthful. The consultant speaks on behalf of the state and he is supposed to be as unbiased as possible. Political communication however, is partial, partisan, and biased because this is the way politicians defend their interests as well as their decisions and opinions. The boundary is invisible, but it is of crucial importance in a sound democracy. "Honest politicians" and their communicators tend not to cross this line. They consider the respectability of the institution above any personal interest (Mazzoleni 2004; Foa 2006).

Spin doctors tend to erase this boundary because they believe that their clients', i.e. politicians' interests are above all else. They believe that in the information age there is no room for ethical concerns, and thus they behave very cynically. Their news management turns into news manipulation, "transparency" becomes deception, "accuracy" distortion, and influencing the news agenda results in hidden propaganda.

2 A hidden distortion of power

Modern spin doctoring began in the 1980s in the United States during the Ronald Reagan administration as a reaction to the superpower of the media. After the Vietnam War and the Watergate scandal, the press was able to intimidate the political world. The media functioned not only as an effective watchdog, but also as the real fourth estate – even at times becoming too influential. Reagan was persuaded that it would have been impossible for any administration to run the country and to pursue public interest if the media had a sort of instant veto power. Reagan then decided – after being advised by clever communication consultants like Michael Deaver – to correct this trend and to reach a more balanced scenario between the media and the political world (Foa 2006).

But spin doctors became too effective. Instead of just balancing, they secretly distorted the news arena to work in their favor. Spin doctors hid their actions because institutions are still the main sources of news. Spin doctoring is the art of twisting facts and turning them around until they appear in the "right" light and until the story has the right "spin" to please those who have commissioned it. Using such extreme methods of PR for governmental and political institutions becomes a very powerful and dangerous tool (Bernays 2004; Bernays 2005).

Spin doctors operate by using a profound knowledge of the mechanisms that govern the information cycle as well as the sophisticated techniques used for the psychological conditioning of the masses. Techniques include, bandwagoning (attempting to persuade the target audience to join in and take the course of action that "everyone else" is taking), appealing to fear, appealing to authority, oversimplification, glittering generalities (emotionally appealing words so closely associated with highly-valued concepts and beliefs that they carry conviction without supporting information or reason), common man (the "plain folks" or "common man" approach attempts to convince the audience that the propagandist's positions reflect the common sense of the people; it is designed to win the confidence of the audience by communicating in the common manner and style of the target audience), scapegoating (the practice of singling out any party for unmerited negative treatment or blame as a scapegoat), stereotyping, obtaining disapproval (this technique is used to persuade a target audience to disapprove of an action or idea by suggesting that the idea is popular with groups hated, feared, or held in contempt by the target audience), etc.

They are clever people who know one of Sun Tzu's lessons in *The Art of War* very well. The lesson says that if you know your enemies and know yourself, you will not be imperiled in a hundred battles; if you do not know your enemies but do know yourself, you will win one and lose one; if you do not know your enemies nor yourself, you will be imperiled in every single battle. There is no doubt that spin doctors know journalists, and they take advantage of their strategic privilege. They know how journalists select news, they know how to feed them with a new story every day. They know how to milk a story, using diversionary tactics like fire breaking or stoking the fire, laundering bad news or "burying" them on days when public attention is absorbed by the Olympics or a soccer championship. They manage expectations. They also manage relationships in the lobby, creating privileged relations with some journalists while intimidating others.

Spin doctors use some of the most advanced public relations tactics designed to exploit the failings of modern media. They know how to influence gatekeeping, how to bypass and exploit high journalistic standards like double-checking and balanced reporting. They have created a culture where the abuse of leaks through unnamed sources is systematic, where an oppressive information's flaw control is normal. They promote messages regardless of factual accuracy and they use well-designed phrases and strategically crafted arguments to distract, deceive and mislead.

3 Spin doctors 2.0

In addition, spin doctors are increasingly leveraging Internet technologies. For instance, as revealed by the Guardian, the U.S. military "is developing software that will let it secretly manipulate social media sites by using fake online personas to influence Internet conversations and spread pro-American propaganda" (Fielding/Cobain 2011).

According to the Guardian, the project has been likened by Web experts to China's attempts to control and restrict free speech on the Internet. "Critics are likely to complain that it will allow the (U.S.) (...) military to create a false consensus in online conversations, crowd out unwelcome opinions and smother commentaries or reports that do not correspond with its own objectives."

This is just one example of how spin might work in the digital age. Overall, Internet misinformation and manipulation seem to be much easier to spread on a day-by-day basis (Mintz 2012), and this comes with no surprise. In an age when practically every computer or smartphone user can publish a blog post or video, and thereby manipulate public opinion, truth becomes a commodity that can be bought, sold, packaged, and ultimately reinvented – as media critic Andrew Keen (2007) argues.

How seemingly easy it is to disseminate false information – not only in the blogosphere, but also in most renowned journalistic media, such as the *ABC News* and *The New York Times* – reveals the self-proclaimed media manipulator Ryan Holiday (2012).

"With a little creative use of the Internet," as Forbes contributor Dave Thier (2012) writes, Holiday has been "quoted in news sources from small blogs to the most reputable outlets." Holiday used *Help a Reporter Out*, a service that connects reporters with sources. Reporters send a query and people who want to contribute to a particular story e-mail back. And Holiday did e-mail back. On several occasions – whether or not he knew anything on the topic. That is the way he lied himself into *The New York Times*. He was quoted in an article on vinyl records. Holiday explained why he liked them – without actually collecting any.

If one wants to understand news and the information society, one must consider manipulators and spin doctors.

Another instrument increasingly used by spin doctors are TV video news releases (VNRs, often referred to as fake TV news by journalists). They have become widespread in the U.S., since Bill Clinton's administration. VNRs are video clips that are indistinguishable from traditional news clips and are sometimes screened unedited by television stations – without the identification of the original producers or sponsors, which in this particular case are government agencies. These are ready-to-use TV news sequences; 90 second stories designed to appear "real," sometimes even with a "real" journalist as the on-air anchor. These VNRs are faked journalism, or public relations. Unfortunately, television viewers can't distinguish between true journalism and such "undercover" PR, which is pure propaganda, produced by and for the government.

4 Spin in a globalized world

Spin doctors are involved in activities that are worrying – particularly concerning secret and intense use of PR companies beyond institutional checks and balances. For example, the U.S. government spends huge amounts of taxpayer money on activities which employ private PR agencies as contractors. Sometimes this is legal and useful, but sometimes these companies are used for hidden political purposes – which is at least puzzling. The Rendon Group, since the beginning of the George H. W. Bush presidency in the late 1980s, has functioned as a "secret power" inside the administration, often without any democratic control. Did Barack Obama change the situation? Rather not, spin doctors and lobbyists still have enormous power.

Spin mainly works in the Anglo-Saxon world (e.g. Jones 2000; Jones 2002), but as we live in a globalized world and in the age of communication, politicians in Continental Europe tend to use spin doctors' techniques as well. The situation may not be as alarming as in the United Kingdom or in the U.S., but indicators in countries like France, Italy, Germany, and Spain can be detected – indicating that spin is more and more widespread. It is a temptation hard to resist, whether governments are center-right or center-left. In fact, spin is neither right- nor left-wing. However, in a democracy, governments are supposed to provide accurate information. Public trust is a value and should not be betrayed.

Are journalists aware of spin doctors' enormous influence? Unfortunately not. This is also part of the failing of the modern media. Nobody teaches journalists about Sun Tzu's lessons. They do not know enough about spin doctors. Spin doctors, however, know journalists.

Some years ago I had a very lively public debate with Klaus Kocks (see Kocks in this book), a very smart and well-known German PR consultant. We disagreed on everything. During the Mainz workshop however, we agreed on at least one thing – he also thinks that journalists' "naiveté" on spin doctoring is exactly what spin doctors dream about. And this of course explains why they usually prevail.

5 A new ethical mission

In conclusion, spin and governments are a very dangerous combination. The widespread use of these misleading techniques has an ultimate result, skepticism towards governmental institutions and a crisis of democracy.

This is a crucial issue. And, in order to preserve their credibility, journalists have a new challenge – even an ethical mission – counter the spin.

References

Bernays, Edward L. *Crystallizing Public Opinion*. New York: Kessinger Publishing, 2004.

Bernays, Edward L. *Propaganda*. New York: Ig Publishing, 2005.

Fielding, Nick/Cobain, Ian. Revealed: US spy operation that manipulates social media. Retrieved on 12th August 2012, from http://www.guardian.co.uk/technology/2011/mar/17/us-spy-operation-social-networks (2011).

Foa, Marcello. *Gli stregoni della notizia*. Milano: Guerini e Associati, 2006.

Holiday, Ryan. *Trust Me, I'm Lying: Confessions of a Media Manipulator*. New York: Portfolio/Penguin, 2012.

Jones, Nicholas. *Sultans of Spin*. London: Orion, 2000.

Jones, Nicholas. *The Control Freaks*. London: Politico's Publishing, 2002.

Keen, Andrew. *The Cult of the Amateur: How Today's Internet is Killing Our Culture*. New York: Doubleday, 2007.

Mazzoleni, Gianpietro. *La comunicazione politica*. Bologna: Il Mulino, 2004.

Mintz, Anne P. (ed.). *Web of Deceit: Misinformation and Manipulation in the Age of Social Media*. Medford: Information Today, 2012.

Thier, Dave. How this guy lied his way into *MSNBC, ABC News, The New York Times* and more. Retrieved on 12th August 2012, from http://www.forbes.com/sites/davidthier/2012/07/18/how-this-guy-lied-his-way-into-msnbc-abc-news-the-new-york-times-and-more (2012).

Part 4: **Search Engines and Social Media**

Christoph Neuberger
Competition or complementarity?

Journalism, social network sites, and news search engines

Abstract: How can the relationship between professional journalism, social network sites, and news search engines be characterized? The results of several empirical studies show that there is no major competition on the part of social network sites and news search engines: People do not lose their consciousness for quality when using the Internet. They see sharp differences between formats and services. Much more important are the complementary connections. Social network sites and news search engines direct significant numbers of people and traffic to websites operated by print and broadcast media. In the opposite direction, journalists use social network sites and search engines for research purposes.

Keywords: journalism, social media, social network sites, news search engines, complementarity

1 Introduction

The relationship between professional journalism, social network sites, and news search engines can be characterized as both competitive and complementary. For *competition* to predominate, the public would have to regard social network sites and news search engines as providing services essentially similar to journalistic websites. The question of whether this is the case was explored in a user survey. The survey's results will be presented in detail in the second section, yet to state a key finding upfront, it is unlikely that journalism is in competition with social network sites and news search engines. Because of that, in the third section *complementary* aspects will be discussed and the question will be answered as to what extent journalistic websites obtain visitors from, or direct visitors to, social network sites and news search engines. In this regard, this section will rely primarily on the results of a study undertaken by the Pew Research Center's Project for Excellence in Journalism that observed Web traffic patterns in the U.S. based on NetView data.

2 Competition between professional journalism, social network sites, and news search engines

What specific features do Internet users associate with different formats and services? This was the research question in a study for which 1,000 German Internet users were surveyed between January and March 2011.[1] The survey participants were recruited in an online panel and were matched to the general distribution of the population with respect to age, gender, and educational level. The survey respondents were asked to evaluate services used at least once every six months (see Table 1) and to checkmark up to three services from a list that embodied a particular feature to the highest degree.

Table 1: Use of services and formats on the Internet in Germany. In %. User survey, January to March 2011. (Source: own data).

	Several times a day	Once a day	Several times a week	Once weekly	Several times a month	Once per month	At least every six months	Less often	Never	I don't know
News search engines (n=1000)	26.5	12.6	19.8	5.9	6.2	3.1	2.0	8.3	13.6	2.0
Video portals (n=1000)	11.2	7.4	20.5	9.7	13.8	9.4	3.3	11.0	12.1	1.7
Wikipedia (n=1000)	7.4	8.0	21.6	10.9	15.1	8.5	4.2	9.9	12.7	1.7
Social network sites (n=1000)	27.2	13.0	12.4	4.2	5.1	3.6	2.2	7.3	23.8	1.1
Portals with news (n=1000)	15.2	15.4	12.9	6.2	8.1	5.6	2.8	12.4	19.6	1.7
Internet services provided by television and radio stations (n=1000)	6.0	7.0	15.6	9.2	11.0	8.3	4.0	16.7	20.1	2.1
Internet services provided by newspapers and magazines (n=1000)	10.3	10.1	14.0	7.7	9.3	5.1	5.3	16.6	20.0	1.5
Weblogs (n=1000)	3.7	4.2	8.1	5.6	6.9	4.4	3.1	15.5	41.2	7.4
Participatory news platforms (n=1000)	4.7	4.3	5.6	3.5	6.4	3.7	2.5	16.9	40.7	11.7
Bookmark collections (n=1000)	2.8	3.1	3.6	2.3	4.1	2.9	2.0	11.0	40.0	28.1
Twitter (n=1000)	2.8	3.0	3.8	2.3	3.7	2.3	1.9	9.4	64.4	6.5

1 The study was supported by Deutscher Fachjournalisten-Verband (DFJV), Berlin. The field phase was conducted by the marketing and media research institute result GmbH, Cologne.

Table 2: Characteristics of Internet services. Choice of the services that embody the enlisted quality criteria to the greatest degree. Up to three services could be selected. Percentage of those surveyed who use a service or format at least once every six months (without the response: "I can't say"). User survey, January to March 2011. (Source: own data).

	Video portals (n=753)	Twitter (n=197)	Social network sites (n=678)	Weblogs (n=359)	Bookmark collections (n=209)	Participatory news platforms (n=307)	News search engines (n=762)	Web services of newspapers and magazines (n=619)	Web services of television and radio stations (n=612)	Portals with news (n=662)	Wikipedia (n=757)
Regular reporting	7.0	15.4	13.9	10.0	7.5	10.5	38.8	63.1	57.5	46.4	10.0
Credibility	7.4	9.9	10.5	7.7	6.6	15.0	30.6	57.4	47.8	37.5	43.0
Broad overview of the news	7.6	7.4	9.9	8.0	6.1	17.2	45.9	56.9	48.7	49.1	17.0
Timeliness	12.4	22.0	26.9	9.3	4.2	14.6	45.0	56.6	50.7	49.3	17.1
Objectivity	4.8	5.0	9.0	8.2	6.6	16.2	31.2	56.5	41.4	34.8	51.2
Issue expertise	8.4	8.9	8.6	11.7	4.7	14.2	29.2	53.0	42.8	32.4	40.3
Issues that are important for everyone	11.2	8.0	14.4	7.9	5.6	18.0	37.3	51.5	45.1	40.0	40.3
Usefulness in everyday life	15.0	11.5	20.3	11.3	7.9	13.6	37.2	47.0	36.3	32.3	54.5
Background information	6.9	12.3	11.1	13.1	8.0	16.8	23.6	44.9	35.8	27.0	41.4
Reader-friendly	17.2	20.4	30.9	20.8	10.3	22.0	25.7	43.8	31.1	31.4	31.6
The author is known	11.8	13.3	18.3	14.2	4.7	12.0	14.2	37.8	24.1	15.0	19.4
Transparency of sources	9.2	7.9	9.6	7.4	7.5	18.8	25.5	36.2	21.0	23.4	46.3
Own research	15.5	12.0	22.3	19.8	6.6	20.4	17.9	34.8	24.1	15.4	34.9
Independent	20.6	17.9	23.2	16.9	10.4	19.7	21.8	26.8	21.1	19.4	34.6
In-depth discussion	19.1	29.8	46.1	32.1	7.1	25.1	12.7	20.3	14.2	18.6	11.0
The author's personal point of view	31.6	27.9	42.1	36.0	8.0	25.1	10.7	16.8	13.5	13.6	22.5

The results were unequivocal: Among the 11 formats and services that were presented as choices, it was the services provided by professional journalists from the *traditional mass media* and the Internet encyclopedia *Wikipedia* that best matched the individual quality criteria (see Table 2). In the opinion of the largest portion of those surveyed, press websites best embodied most of the features. Although Wikipedia and television and radio station websites did receive fewer top votes, they were still rated significantly higher than all other services.

With the exception of Wikipedia, social media was scarcely associated with journalistic features. Only in relation to questions about the author's personal viewpoint and about in-depth discussions did social network sites (and weblogs) receive endorsement by the largest portion of those surveyed. Neither of these features is a central characteristic of journalism, however.

In the case of Wikipedia, it is worth noting that the service does indeed meet the criteria for journalistic quality, but it was primarily valued for its usefulness in daily life and the encyclopedic information it provides. However, Wikipedia has none of the identifying features of journalism, such as regular reporting, timeliness, and a broad overview of the news. The fact that Wikipedia did better than press websites with respect to independence as well as transparency of sources reveals that the collaborative contributions of lay persons can outperform professional press output in terms of core journalistic quality criteria, at least in the opinion of the users. News search engines like *Google News* performed better than social media and were valued primarily for the broad overview of the news they provide and their timeliness.

Besides these features, the respondents were also asked about their reasons for using the sites. In a list of 13 gratification items, specific details about information behavior on the Internet were considered. The list differentiated between the passive reception of a pre-produced menu of subjects (surveillance), active search (guidance), accidental reception (serendipity), as well as the user's (meta-)orientation regarding current information services on the Internet itself. Besides current news information, which is the central focus of journalism, the list also included the acquisition of basic background knowledge, advice, entertainment, engagement in discussion, and the cultivation of relationships – and thus included areas at the outskirts and edges of journalism.

In relation to the user attribution of features to media formats, the results concerning users' reasons for using the sites turned out quite differently (see Table 3):

- Users also prefer to engage with legacy media on the Internet if they want to obtain an *overview* about the region, Germany, or foreign news. According to users, professional journalism also defines the topics one should be familiar with as a citizen. This corresponds to the traditional *gatekeeper* and *agenda-setting* role ascribed to professional journalism.
- However, if users want to be personally active and are *searching* for specific information or reports about *special areas of interest*, then they prefer news search engines and the online encyclopedia Wikipedia.
- News search engines and portals with news are especially well suited for *orienting users on the Internet* and the *accidental discovery* of interesting subjects.

- The strength of social media lies in *communicative gratification*: Social network sites and weblogs are appropriate sites for discussion, whereas for cultivating relationships, users prefer social network sites and Twitter.
- People seeking *entertainment* prefer social network sites and video portals.

Table 3: Reasons for using Internet services. Choice of services that are used most often for this purpose. Up to three entries are possible. Percentage of respondents who use a service or format at least once every six months (without the response: "No opinion"). User survey, January to March 2011. (Source: own data).

	Video portals (n=753)	Twitter (n=197)	Social network sites (n=678)	Weblogs (n=359)	Bookmark collections (n=209)	Participatory news platforms (n=307)	News search engines (n=762)	Web services from newspapers and magazines (n=619)	Web services from television and radio (n=612)	Portals with news (n=662)	Wikipedia (n=757)
To get an overview of current events in Germany	4.2	8.9	10.3	7.4	7.0	10.1	34.7	59.6	44.4	40.4	7.2
To get an overview of current events in my region	3.4	5.9	15.7	8.0	5.6	9.5	27.1	54.0	28.2	20.8	4.5
To get an overview of current events abroad	4.4	6.9	8.2	8.0	3.8	12.1	34.1	48.2	36.5	36.4	6.3
To find background information about current news topics	5.1	6.0	7.4	7.7	8.0	14.9	32.3	43.4	27.8	25.4	34.8
To be able to converse about important issues	9.4	13.4	26.5	13.5	6.1	13.7	19.2	38.6	27.5	22.6	20.9
To find out about specific subjects of interest to me	9.9	9.5	10.6	13.3	7.5	9.8	40.9	32.1	19.5	20.2	57.0
To learn about things that are useful in daily life	11.2	9.4	16.8	16.2	6.6	13.4	32.8	32.0	22.7	26.4	39.4
To stumble upon interesting stories	21.2	15.8	22.2	14.3	8.5	14.0	32.6	29.6	23.5	33.1	19.8
To learn which Internet sites have information about current issues	5.5	6.9	11.2	12.1	12.1	7.6	44.9	25.4	16.9	25.6	11.5
To search for specific information	7.6	8.5	8.9	9.1	9.3	9.8	50.6	23.8	16.4	20.3	58.2
To entertain myself	47.1	18.0	52.0	13.8	9.4	5.7	12.7	18.1	26.5	16.5	8.3
To have discussions with other people	9.6	19.9	46.4	22.5	4.3	12.5	8.1	13.1	9.8	10.7	5.7
To cultivate my social relationships	8.6	25.8	67.4	9.6	4.2	9.5	6.7	8.4	5.0	10.7	4.6

Social media outlets are largely unimportant as a news source (i.e. for an over-view of current events). However, social network sites do at least permit people to find out which subjects are important for conversation. Presumably, this is due to the fact that social network sites spread the word on subjects that are important in the media.

In summary, the results of the user survey show that people do not lose their consciousness for journalistic quality when using the Internet. Internet users see sharp differences between formats and services and in most categories they prefer professional journalism. So it is unlikely that competition exists between journalism, social media, and news search engines.

3 Complementarity: The forwarding of users and journalistic research

The dissemination of news story-links on social network sites and other social media outlets has grown increasingly important for journalistic websites, as shown by several recent studies. In their study, *Understanding the participatory news consumer*, Purcell et al. (2010) conducted a representative telephone survey in December 2009 and January 2010. The survey found that 17 percent of Internet users in the U.S. posted links or thoughts about news stories on social network sites such as Facebook (Purcell et al. 2010: 61). An additional 3 percent posted or retweeted a link to an online news story or a blog entry on Twitter.

Such content recommendations provide a form of orientation, and if users comment on news stories, they become a type of pre-filter for other users. In this way, they help to draw attention to a story as well as enhance or diminish confidence in the quality of journalistic websites. What impact do such link recommendations have on Web traffic flows? How much do they contribute to overall traffic of a particular website?

Olmstead, Mitchell and Rosenstiel (2011) used NewView data from Nielsen to examine how users get to the top 25 news websites in the U.S. They based their analysis on data from the first three quarters of 2010. Nielsen records traffic from a referral site when at least five users were linked from it. Data is available for 21 of the 25 top sites.

- On average, 30 percent of the traffic to the top sites came through Google (Olmstead/Mitchell/Rosenstiel 2011: 7 f.), referring here to all Google services. For 17 of the 21 sites, Google was the largest supplier of traffic, and it was the second largest for the four other services.

- Of course, the majority of *Google News* users did not limit themselves to reading the headlines but also clicked on news stories. However, only a few large news websites benefited from this, especially *The New York Times* (15 percent of all users of the *NYT* on the Internet), *CNN* (14 percent) and *ABC News* (14 percent). These three services accounted for 69 percent of all *Google News* users who followed a link to another site (Olmstead/Mitchell/Rosenstiel 2011: 22). *Google News* promotes the concentration of attention on the Internet.
- Facebook was the second or third most important access site for five of the services. Sites that benefited the most from Facebook included the *Huffington Post* (8 percent), *CNN* (7 percent) and *The New York Times* (6 percent) (Olmstead/Mitchell/Rosenstiel 2011: 10, 24).
- Surprisingly, the *Drudge Report* was even more significant than Facebook: Matt Drudge's online magazine, which became well-known during the Lewinsky scandal in 1998, was the second or third most important user generator in 12 cases (Olmstead/Mitchell/Rosenstiel 2011: 11).
- However, Twitter proved to be quite insignificant (Olmstead/Mitchell/Rosenstiel 2011: 12).

How often do users go from a news website to a sharing site and share what they have read (Olmstead/Mitchell/Rosenstiel 2011: 19, 24)? Here, it was mostly Facebook that dominated; up to 7 percent of users (*Yahoo News*) shared news items on Facebook (6 percent for *CNN*; 4 percent for *ABC News*, *Fox News* and *The Huffington Post*; 3 percent for *The New York Times* and *The Washington Post*).

According to an evaluation conducted by Comscore in Germany (Schmidt 2011), Google Sites and *Google News* lost market share due to traffic being directed to media websites. By contrast, Facebook witnessed a traffic increase. In February 2011, *bild.de* obtained 10 percent of its visits through Facebook (February 2010: 2 percent). An increase in visits could be shown for *spiegel.de* (2011: 4 percent), *focus.de* (3 percent) and *faz.net* (3 percent). Even though the percentage of users who visit a professional journalism website via Facebook is still in the single-digit realm, it appears that recommendations from one's circle of friends and acquaintances are also gaining importance as a pre-filter in Germany.

The possibilities for disseminating and reading news stories over social network sites were documented in a more differentiated way in a non-representative online survey conducted by Wladarsch (2010). This survey showed that 49 percent of social network site users had commented on media contributions or clicked the *like* button. Of the users surveyed, 34 percent had forwarded media articles themselves or transferred them from a media website to a social network site (Wladarsch 2010: 141).

The presented data show that the forwarding of news stories through social network sites drives traffic to journalistic websites. Yet news search engines also make a significant contribution to directing users to journalistic websites. This is not the only complementary relationship, in the opposite direction journalists also use search engines and social network sites for research purposes.

This is shown, for example, in the findings of our newsroom survey, which collected data from editors-in-chief of Internet news departments in May and June 2010. Survey participants were members of German news departments from several superregional media organizations (Neuberger/vom Hofe/Nuernbergk 2011). In the case of daily newspapers, all titles ("publizistische Einheiten") were analyzed, including regional publications. Altogether 70 newsroom directors took part in the survey, reflecting an overall return rate of 45 percent. The study was sponsored by the Media Authority of North-Rhine Westphalia (Landesanstalt für Medien Nordrhein-Westfalen, LfM). How often do staff members use computer-based research methods (see Table 4)? While search engines and Web catalogues predominate, social network sites, together with other social media, are still of only relatively small significance.

Table 4: Use of computer-based research methods by newsroom staff. In %, n = 60, Newsroom survey, 2010. (Source: Neuberger/vom Hofe/Nuernbergk 2011: 60).

How often do staff members use the following computer-based research methods?	Frequently	Rarely	Never
Search engines and Web catalogues	98.3	1.7	0
Online-services provided by journalistic media	98.3	1.7	0
E-mail	95.0	5.0	0
Internal house archives	83.3	15.0	1.7
Databases and archives	80.0	18.3	1.7
Weblogs	30.0	56.7	13.3
Social network sites	21.7	73.3	5.0
"Social news" services	13.3	75.0	11.7
Twitter	11.7	86.7	1.7
User platforms	8.3	73.3	18.3

Three-step scale. The response "other" is not included. The response "I cannot comment on this matter" was not included in the analysis.

4 Satisfied with reading signposts

The considerations presented thus far could lead to the conclusion that the world of journalism is in order, as there is no major competition from social network sites and news search engines. Indeed, quite the opposite is true. Social network sites and news search engines direct significant traffic to websites operated by print and broadcast media. Furthermore, newsroom staff uses social network sites and news search engines as a research tool. This harmonious interplay has been portrayed by Bowman and Willis (2003: 12) in the form of a new media eco-system model.

Table 5: Information behavior and evaluation on the Internet (1 = agree completely, 5 = disagree completely, without the response "I can't say"). (Source: own data).

	Average	Agree completely / Agree somewhat, in %
Because I can also obtain adequate information on other sites for free, I am not prepared to pay for journalism on the Internet. (n=938)	2.3	61.4
I come across many news stories on the Internet by chance, so I don't have to search for them specifically. (n=942)	2.7	42.4
In order to be aware of the important topics of the day, I don't have to visit any media websites, since there is also information about them presented and discussed on other sites. (n=918)	2.8	36.3
I find it difficult to evaluate the quality of information on the Internet. (n=956)	3.1	27.6
It's hard for me to distinguish what is journalism on the Internet and what is not. (n=950)	3.1	25.3
When I visit Google News, I only read the headlines and don't call up the whole article. (n=838)	3.1	23.4
Recommended links that I obtain from friends in social network sites are more valuable than search machine results. (n=862)	3.3	20.2
I obtain many news items related to politics and business directly from political party and company websites. Therefore, I don't have to rely on journalism. (n=908)	3.9	13.0
On the Internet, I have the sense that I can't find what I am looking for. (n=950)	3.7	12.2

Yet as we all know, this image of an *ideal world* is misleading. The Internet has shaken the very economic foundations of professional journalism. It is clear however, that professional media providers must bear a substantial share of the responsibility for this: The profusion of choices, their omnipresence, their availability free of charge, and the interchangeability of services has diminished brand loyalty and reduced the audiences' willingness to pay for services. This is confirmed by the survey of Internet users cited at the beginning of this report (see Table 5).

The study also shows that social network sites and news search engines contribute to the superficial consumption of news stories:

– Nearly one quarter of respondents were content to read only the headlines in *Google News* and did not access the entire news article.
– In addition, important subjects were also passed along through *social network sites*. In any case, many users still feel sufficiently well informed to be able to talk about important subjects (see Table 3).

Clearly, many users on the Internet are satisfied with merely reading signposts – instead of clicking on headlines with the aim of reading the entire news story. For such users, the signpost is the aim.

References

Bowman, Shayne/Willis, Chris. We media. How audiences are shaping the future of news and information. *NDN*. Retrieved on 12th November 2011, from http://www.hypergene.net/wemedia/download/we_media.pdf (2003).

Olmstead, Kenny/Mitchell, Amy/Rosenstiel, Tom. Navigating news online: Where people go, how they get there and what lures them away. Retrieved on 12th November 2011, from http://www.journalism.org/sites/journalism.org/files/NIELSEN%20STUDY%20-%20Copy.pdf (2011).

Neuberger, Christoph/vom Hofe, Hanna Jo/Nuernbergk, Christian. Twitter und Journalismus. Der Einfluss des „Social Web" auf die Nachrichten. LfM-Dokumentation 38. 3rd ed. Düsseldorf: Landesanstalt für Medien Nordrhein-Westfalen (LfM). Retrieved on 3rd November 2011, from http://lfmpublikationen.lfm-nrw.de/catalog/downloadproducts/L043_Band_38_Twitter.pdf (2011).

Purcell, Kristen/Rainie, Lee/Mitchell, Amy/Rosenstiel, Tom/Olmstead, Kenny. Understanding the participatory news consumer. How Internet and cell phone users have turned news into a social experience. Retrieved on 12th November 2011, from http://www.journalism.org/sites/journalism.org/files/Participatory_News_Consumer.pdf (2010).

Schmidt, Holger. Trafficlieferanten der Medien: Facebook gewinnt, Google verliert. *faz.net*. Netzökonom. Retrieved on 12th November 2011, from http://faz-community.faz.net/104216/print.aspx (2011).

Wladarsch, Jennifer. Auf der Spur der Massenmedien in sozialen Onlinenetzwerken. Wie und warum Internetnutzer massenmediale Inhalte in sozialen Onlinenetzwerken nutzen. Unpublished Master's Thesis Ludwig-Maximilians-University Munich, 2010.

Peter Laufer
Crowdsourcing is nothing new

Professional journalists are still needed to determine what is news

Abstract: News organizations must innovate and adapt to developments of the converging media world. In this respect, social media can serve as a distribution device for journalistic content and as a source for reporting. However, crowdsourcing is not merely a phenomenon of the digital age. Journalists have always harnessed the audience as a source of news – although by using other technologies and calling this crowdsourced material news tips. Therefore, there is no need for journalists to rethink their reporting. Moreover, it would be a mistake to suggest that the audience should dictate the contents of a newspaper or a broadcast. It is the job of professional journalists to determine what is news.

Keywords: crowdsourcing, citizen journalism, news consumer, news tips, credibility

1 The healthy diet of credible news sources

Is citizen journalism, swarm intelligence, and crowdsourcing forcing journalists to rethink their reporting? For me, the answer is a resounding "No."

Of course journalists need to embrace the latest technologies and opportunities that new tools present to them. Social media as a source can help journalists learn more about what's happening around the world and thus influence their reporting, their choice of stories, and their content. Social media can be harnessed as a distribution device – a powerful asset for adding mass to media.

But social media as a news source is simply another term for what journalists used to call news tips. And social media as a distribution device is simply another term for what used to be referred to as a newspaper delivery truck.

And above all, Internet chatter cannot replace the investigative work of a traditional news reporter in much the same way that gossip at a café or a barbershop cannot replace a healthy diet of credible news sources.

2 Citizen journalist vs. citizen dentist

It is a fool's mission to suggest that readers should dictate the contents of a news-paper, a broadcast, or other forms of media. A journalist's job is to report the news by first figuring out what is going on in the world, then determining what they believe their audiences should know, and finally establishing what should be included in their reports before they are delivered.

Especially in today's era of instant communication, many readers do not necessarily need professional journalism coverage to find what they deem as important information. A journalist's value – to use the overused term now in the vernacular – is as *curator* of worldwide current events, followed by their role as reporter.

Two examples from leaders of industries other than journalism reinforce my insistence that citizen journalism is an oxymoron unless that citizen happens to be a professional journalist. Journalist Andrew Stroehlein from the International Crisis Group in Brussels explained that, "citizen journalism is like citizen dentistry" (Stroehlein 2010). Do citizens really want to pull their own teeth?

Even the term *citizen journalism* is cause for concern. Attaching the modifier *citizen* suggests purity. There is no reason to anoint journalism-like offerings from the public with de facto credibility based on nothing but the fact that they supposedly came from the public. Consider two possibilities. First, think about how easy it is for an agent provocateur to foment false or misleading reports that hide behind the moniker of citizen journalism. And second, think about the lack of context that comes with a mobile phone video clip posted by a *citizen journalist* on YouTube.

3 Two examples from outside the fourth estate

My first example that contradicts crowdsourcing comes from the big box store, IKEA.

Whatever you may think about the imperialist nature of IKEA, not to mention the ubiquitous nature of their designs and wares, you cannot argue with the store's success. IKEA doesn't crowdsource or make use of citizen furniture designers, the company instead understands its role is to develop products and bring them to consumers. In a *New Yorker* magazine profile of the company, the manager of the IKEA in Malmö, Sweden, Martin Albrecht, said IKEA wants to "show (customers) how to live!" And about the products on offer, he added: "If you don't show the customer the function, the customer won't understand it" (Collins 2011).

I am not equating the vital work of journalism with peddling housewares, but the philosophy transfers. We cannot expect journalists to teach the audience much about how to live if the news coverage comes back to them in a closed circle: crowdsourced from the public, augmented by journalists, and regurgitated back to them. A journalist's job is to determine what is news, and then – to paraphrase IKEA – explain it to the audience, because if they don't, the audience won't understand it.

Arrogant? You bet. That defines a journalist's job, and a taste of elitism is in order here.

The other anti-crowdsourcing example I found was buried deep in *The New York Times* obituary for Steve Jobs. When he was asked what market research went into the development of the iPad, his answer was quick. "None," he said. "It's not the consumers' job to know what they want" (Markoff 2011).

Exactly. And to add a word to the Jobs quote, it is not the *news* consumers' job to know what they want.

4 Crowdsourcing in proper perspective

Not that crowdsourcing should be tossed out as a tool. After all, it is nothing new. It was October 1989 and I was working at radio station *KCBS* in San Francisco, announcing the news from our studios on the 32nd floor of a downtown skyscraper. Just hours before, northern California had been hit by a killer earthquake and overworked first responders were unable to determine the extent of the deaths, injuries, and damage. At the radio station, as our building swayed from the aftershocks, we solicited telephone calls from our audience, asking them on air where they were and what they knew.

We made it clear to our audience that these were unverified reports and began mapping the earthquake and its effects – even realizing, after a report of a sharp aftershock fifty miles south of us, that we were going to be swaying ourselves again in moments. A rare example of crowdsourcing emerged which allowed us to not just report events, but to forecast the news for our audience. This system of crowdsourcing helped us decide where to send our reporters long before the term was coined. The eyewitness reports were detailed, graphic, and dramatic. We chose to air them live and unverified because we decided it was unlikely anyone would fabricate a story at such a time of common duress, and because we aired so many reports, even if one were untrue, the many others would dilute its erroneous effects.

The same approach holds for Internet-based crowdsourcing. However, by making use of news tips, or printing letters to the editor – another form of old school crowdsourcing by a different name – journalists are not pardoned from their traditional oversight responsibilities.

5 The customer is not the editor

In October 2011, *The Guardian* in London announced that it would start sharing a selection of its news list with its readers, a list identifying the stories the paper expects to cover each day. Afterwards readers would provide input about what stories to cover and prioritize. The paper's national news editor Dan Roberts was quoted by *MediaBistro* (Zak 2011) as musing, "What if readers were able to help news desks work out which stories were worth investing precious reporting resources in?" If that serves your readership, Mr. Roberts, then perhaps the paper doesn't need you – as assignment editor – to come to work. But your job is crucial to a well-informed electorate. And your training and experience should make you want to do it *without* interference from the chorus.

"Eventually," said Roberts, "I suspect, a generation gorged on reality television and social media might demand this from all sorts of media companies."

What a sorry prognosis. Imagine your neighbors deciding not just your favorite newspaper's headlines, but also what music you'll hear on the radio, what movie will play in your hometown theater, maybe even what your preacher will say from the pulpit.

News reporting requires a filter and incoming information needs to be curated. Factoids, propaganda, and other bits and pieces of a news story need to be fact checked and vetted, and claims and counterclaims must be investigated. A professional observer should report events from the vantage point of an independent eyewitness, in order to ensure balance and credibility. Reporters and editors should enterprise new news stories. If we jettison the journalist, much of the information being generated is just noise without context. I'm not contradicting the idea that we all can be journalists. We all *can* be journalists, but everyone in a crowd is not a journalist. Reports from the crowd are news tips, not news reports. If we want credible journalism, we need professional journalists.

6 Crowdsourcing as chaos and riot

Consider the important journalistic role of a book or a restaurant reviewer for example. When self-selected anonymous reviewers show up on Amazon and claim a book is not worth reading, why should we pay attention to these opinions? We do not know if the reviews were written by an enemy of the author to trash the book. Why should we waste time reading a review by a writer unknown to us, and unaffiliated with any organization that would hold the reviewer accountable? The same is true with amateur restaurant reviews, and other anonymous chatter that fills Internet comment pages.

Crowdsourced? Sure. Worth reading? Maybe for its potential entertainment value, but certainly not as credible reporting.

7 Journalism – the world's oldest profession

Even the Occupy Wall Street demonstrators in New York – a group that thrived on being connected through social media – understood the value of harnessing journalism professionals to report the news. Since September 2011 the occupiers published several print editions of their signature publication, *The Occupied Wall Street Journal*. The first press run of 50,000 issues was gobbled up by anxious readers, according to the paper's staff. Former *New York Times* reporter and Pulitzer Prize winner Chris Hedges authored the lead story in its second edition. The paper's editor, Jed Brandt, explained to the *Wall Street Journal* that a newspaper designed to chronicle the occupation was appealing because a newspaper is "so old, it's new" (Firger 2011). Note that there was an editor at the paper making editorial decisions.

Not to say that professional journalists are flawless. Judith Miller reported in *The New York Times* that Saddam Hussein stockpiled weapons of mass destruction. Jayson Blair was a plagiarist in the same paper's pages. But those are exceptions. There are bad dentists who pull out the wrong teeth, too.

Imperfect professionals don't change this reality. We do not need to know the moment-by-moment thoughts and activities of an endless parade of Facebook *friends*, nor they ours. Nor should we consider random crowdsourcing of potential news stories as a threat to traditional news reporting. It is inane to assume such chatter should or would replace professional journalism.

Noise is not news.

References

Collins, Lauren. House perfect. Is the IKEA ethos comfy or creepy? *The New Yorker*, 3rd October 2011: 54–59.

Firger, Jessica. Protesters' newspaper occupies a familiar name. Retrieved on 14th May 2012, from http://blogs.wsj.com/metropolis/2011/10/04/protesters-newspaper-occupies-a-familiar-name (2011).

Markoff, John. Apple's visionary redefined digital age. Retrieved on 14th May 2012, from http://www.nytimes.com/2011/10/06/business/steve-jobs-of-apple-dies-at-56.html?_r=1&pagewanted=all (2011).

Stroehlein, Andrew. PAX: A new idea in conflict prevention? Retrieved on 14th May 2012, from http://www.andrewstroehlein.com/2010/05/pax-new-idea-in-conflict-prevention.html (2010).

Zak, Elana. The Guardian shares newslists with readers. Retrieved on 14th May 2012, from http://www.mediabistro.com/10000words/new-experiment-lets-readers-influence-editorial-decision-making-process-at-the-guardian_b7513 (2011).

Oliver Quiring
Journalists must rethink their roles

Social media – an opportunity for newsrooms comes with obligation

Abstract: Crowdsourcing is contingent on users who are willing to spend time and effort to create news or some type of user-generated content. However according to research, few users are actively contributing to the content pool on the Internet. Users comment, share, and tag – but few are producing their own original news material. Nevertheless, there are at least three major reasons why journalists should rethink their roles and should begin caring about active users and user participation in the Web 2.0: a) Social media changes the process of news diffusion, b) users are already dictating the content of media products, c) and the wide diversity of raw news material on the Internet needs to be carefully curated by professional, independent journalists.

Keywords: crowdsourcing, citizen journalism, diffusion of news, balance of power, quality of news

1 The journalistic and economic perspective

In the context of this article, the terms *citizen journalism, crowdsourcing* and *swarm intelligence* all refer to the ways in which users contribute to the media supply. However, the concepts behind these terms still lack a clear definition.

From a journalistic perspective, citizen journalism serves as an umbrella term for different types of user participation in the process of generating (socially relevant) news. Originally the idea of citizen or so-called *grassroots* journalism (Gillmor 2006) drew attention to the important democratic functions of news reporting and was considered a source of competition for professional journalism (Fröhlich/Quiring/Engesser 2012 forthcoming). Even in 2008, Paulussen et al. still insist that in citizen journalism, "the news making process is completely out of the hands of journalists and is handed over to the people who have become both producers and users of the news" (Paulussen et al. 2008: 267). New forms of publication on the Internet, such as weblogs or news platforms that rely exclusively on so-called citizen reporters, are a perfect outlet for such activities. However, not all of the offers coined as citizen journalism are able to provide high standards in terms of their contribution to democratic societies (Engesser 2008: 63). For example, some forms of citizen journalism simply inform hyper-local communities of what's going on in the neighborhood (e.g. Fröhlich/Quiring/Engesser 2012 forthcoming).

However, reducing the perspective of this article to the narrow meaning of the term citizen journalism (i.e. as strict competition to professional journalism) would give away large parts of the theoretical and empirical potential of the concept of *user participation*. In a recent article, Neuberger and Nuernbergk (2010) mention three different forms of relations between traditional journalism and active users: competition, complementarity and integration. While competition and complementarity take place in different media outlets for professional and participatory journalists, integration provides the opportunity to present professional and user-generated content on one platform. There are many different ways in which this integration can be accomplished, for example by allowing users to send photos and to comment on news stories, and by displaying *most read articles* lists via aggregating data on user behavior. There are many different motives for users (e.g. impression management) and media outlets (e.g. reduction of cost) to seek such an integration (for an overview: Schweiger/Quiring 2005; Fröhlich/Quiring/Engesser 2012 forthcoming).

Broadening the perspective to these kinds of media products offers the possibility to create a theoretical bridge towards the second key concept of this paper. From an economic perspective, crowdsourcing refers to "technological advances in everything from product design software to digital video cameras (...) (that) (...) are breaking down the cost barriers that once separated amateurs from professionals" (Howe 2006). Transferred to the journalistic sphere, crowdsourcing enables former amateurs to create professional-looking media products. It also enables professional media entrepreneurs to integrate user-generated content into their products (Schweiger/Quiring 2005). From an economic perspective this kind of content generation is attractive because of its potential to reduce the cost of news production in times of fierce competition on the advertising market. Drawing on Howe's crowdsourcing ideas, devoted citizens would be able to produce news at a very low cost or for free via ICT technologies.

Theoretically, a random crowd of users on the Internet should provide *swarm intelligence* (an important prerequisite for sophisticated contributions to news making). Swarm intelligence theoretically should ensure quality. Although this alternative way of news production appears to be economically attractive, there are still some decisive questions which have yet to be addressed: Are there enough users willing to work for free? Do users actually create something useful or is the result of crowdsourcing more or less swamp intelligence, meaning everyone dumps information in a pool and other people have to find the precious pieces? Is the idea of crowdsourcing suitable for any kind of project? And finally, do professional journalists have to rethink their roles as a consequence of citizen journalism and crowdsourcing?

2 Journalism's role in the age of sharing

Peter Laufer's answer to the last question was a resounding "no" (see Laufer in this book). Although it is doubtful whether the simple causal chain presented in chapter 1 (citizen journalists, i.e. users willing to participate → crowdsourcing, i.e. giving them a common platform → swarm intelligence, i.e. creating useful news) indeed works, journalists nevertheless should rethink their roles. Referring to Laufer's different hypotheses, I will try to shed light on four sets of arguments: the number of users willing to take an active role, the distribution of media products, the impact of journalists and citizen journalists and aspects of quality.

2.1 Quantity

Crowdsourcing as a tool needs users who are willing to spend time and effort on creating news or at least some kind of user-generated content. Data on the German online population reveals that only a small amount of users are actively contributing to the content pool on the Internet (see Figure 1). Looking at the most important Web 2.0 services, only a small percentage of users produce or upload content, and even then content is often unoriginal or shared. There is another service which is not shown in Figure 1 but which is important: According to the same study a great deal of content is created on social networking sites, but not for public use (Busemann/Gscheidle 2011).

Furthermore, a study published by the Pew Research Center states that user participation in the U.S. "comes more through sharing than through contributing news themselves" (Purcell et al. 2010: 4). The results of a 2009 telephone survey, which used a sample of 2,259 adults aged 18 and older, showed that it is quite common to comment on a news story (25 percent of the sample), to post a link on social networking sites (17 percent), to tag content (11 percent), but that only 9 percent of those surveyed created their own original news material or opinion piece (Purcell et al. 2010: 4).

Internationally, well-known citizen journalism projects like the South Korean website *OhmyNews*, changed their scope from providing news to providing news about citizen journalism. Even the reach of most blogs – a service formerly praised for its democratization of media – is limited to a rather small audience.

Apart from a few (nowadays often institutionalized) A-list blogs, there is a large number of (in many cases very private and strictly non-political) blogs along the so-called "long tail" (Schmidt 2006; Schmidt 2009) that almost no one reads. So why should journalists care?

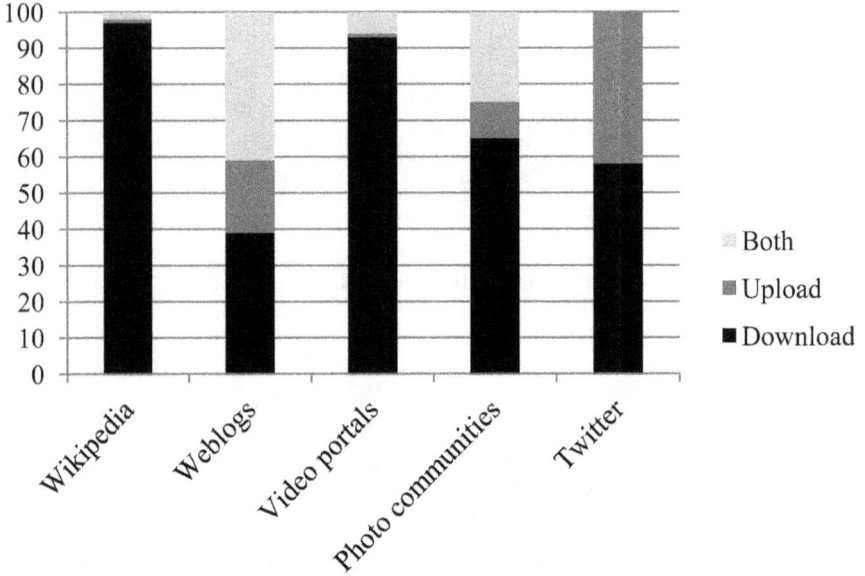

Figure 1: Preferred way of usage of Web 2.0 services. (Source: data from Busemann/Gscheidle 2011; ARD-ZDF-Onlinestudie, German-speaking population aged 14 and more).

2.2 Distribution

Laufer also argues that journalists do not need to rethink their roles, and while social media is "a powerful asset for adding mass to media," it's simply "another term for what used to be referred to as a newspaper delivery truck" (see Laufer in this book). In this line of reasoning, social media would simply add to the reach of news.

Referring to Laufer's delivery truck imagery and focusing on social media, one could say that we deal with a truck that simultaneously travels on roads (e.g. text), rail (e.g. photo) and in the air (e.g. video); is able to create its own routes (e.g. by users who are commenting and hyperlinking to other media outlets); and sometimes operates the driver (e.g. by providing feedback and sometimes forcing corrections, see Figure 2). To state it less metaphorically, social media adds to the reach of news but it also entirely changes the process of news diffusion and production. News diffusion cannot simply be reduced to reaching certain audiences (for an overview on diffusion theory: Rogers 2003). The processes of news diffusion are often complicated by the unpredictable impact of users. Users who are able to share news content and to directly and visibly comment on news stories are gaining influence, making things much more complicated for news producers because journalists are forced to manage user participation (Ziegele/ Quiring 2011). Therefore journalism training must change in order to handle different formats of communication simultaneously, in the same way it has already changed with academic courses on cross-media and online journalism (e.g. Meier 2011).

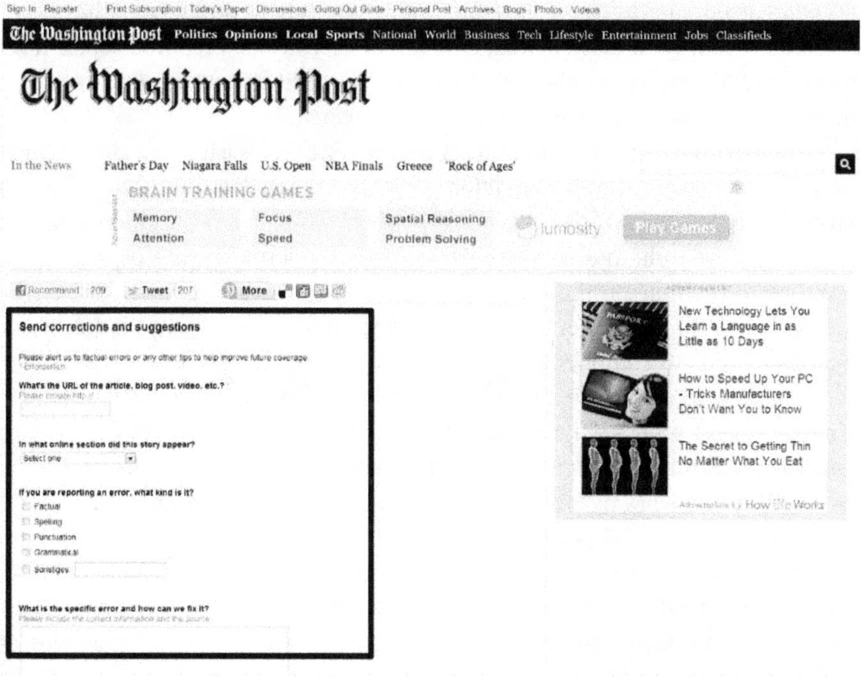

Figure 2: Screenshot of the *Washington Post* Website – "corrections and suggestions."

2.3 Balance of power

This is where Laufer's second key message becomes relevant. He states that, it "is a fool's mission to suggest that readers should dictate the contents of a newspaper, a broadcast, or other forms of media (...), it is not the news consumers' job to know what they want. (...) Imagine your neighbors deciding not just your favorite newspaper's headlines, but also what music you'll hear on the radio (...)" (see Laufer in this book). This means that users should not have the power to dictate the content of media products. However a close inspection of media services reveals that users already do.

A well-known example of user-controlled content is radio, where the number of record sales dictates the playlists. Similar processes apply for news. While the programming of TV stations is already chosen based on media usage and consumer preference research, online media outlets choose news stories based on content popularity (see Figure 3). Complementary to ordinary online usage research, editors can learn a great deal about their users' preferred topics from these lists. It is more than plausible to suggest that editors base content decisions on audience preference in order to sell their product.

In this context it seems important to answer one decisive question: Which users interfere with news production? In some cases this question is easy to answer. In the last example, the "most popular news"-list is based on a simple aggregation formula and technological transformation of reader behavior in a given period of time on the respective website. In this case users only alter the salience of certain topics for other users and provide more or less an indicator for agenda setting (McCombs 2008).

So far, we do not have a clear-cut picture of users who provide journalistic comments and produce their own news stories. Moreover, the composition of active users will inevitably vary between different media services. A small study as well as an overview by Fröhlich, Quiring and Engesser (2012 forthcoming) indicates that a substantial portion of users who commented on or produced news stories for a local German news platform, worked either as professional journalists or in the media business. Therefore, at least some users might know how to provide professional content. They also regarded themselves to a much lesser degree as *neutral mediators* of information as opposed to professional journalists and they selected topics based on their own preferences rather than professional criteria. But the results of the study are only a glimpse of the real phenomenon and the degree to which these results are generally valid is still unknown.

Figure 3: Screenshot of the *Washington Post* Website – "most popular news."

To make a long story short, there is little systematic data on active users but there is data on editorial reactions. According to a study conducted by Neuberger and Nuernbergk (2010), German editors care about active users and are beginning to rethink their roles. First of all, editors do not perceive a major threat to the reader market of traditional journalism and do not fear competition from social media (Neuberger/Nuernbergk 2010: 326). Most editors regard social media as a *complement* to traditional journalism, which not only helps to promote inspired news coverage, but also serves as an agenda-setter in extraordinary situations like the Arab Spring (Neuberger/Nuernbergk 2010: 327). And editors clearly answer the question whom they perceive to be in power. Newsrooms started to *implement* applications for active users and therefore exploit the content and data users deliver but they keep tight control of the content published (Neuberger/Nuernbergk 2010: 329).

2.4 Quality

Besides economic reasons, professional journalists try to control the news production process because they believe that they – and only they – are able to
ensure a high quality of news. Laufer clearly refers to the quality dimension of the
discussion when he states, a "journalist's value (...) is as curator of worldwide
current events (...). (...) News reporting requires a filter (...), (...) pieces of a
news story need to be fact checked and vetted and claims and counterclaims must
be investigated. A professional observer should report events from the vantage
point of an independent eyewitness." In Laufer's opinion, "social media as a
news source is simply another term for what journalists used to call news tips"
(see Laufer in this book).

The discussion of online news quality, specifically the quality of user contributions, is far from new and opinions are mixed on the ability of users to provide
well-constructed news items (for an overview: Beck/Schweiger/Wirth 2004).
Therefore, the question whether citizen journalists are providing *swarm intelligence* and whether they are able to provide better, worse, or just different news
items still remains to be answered empirically. But even if we accept that the true
value of professional journalism lies in the curation of news – the selection, the
verification, the balancing of arguments and the independence of journalists –
social media is more than simply a source for news tips. Social media is an opportunity that comes with obligation. Social media indeed offers seemingly endless
raw content, which can be exploited by professional journalists. Nevertheless,
journalists will have to develop professional rules for how to deal with this opportunity. At the moment it looks as if there are no clear standards.

Figure 4 shows a typical example of the unsuccessful exploitation of raw
material. The *Associated Press (AP)*, much like several other media outlets, published the respective video via YouTube. *AP* labels this material as *raw material*
and indicates that the news agency is not able to verify the content. Unfortunately
AP does not realize that it does not exclusively transfer news items and raw material to professional journalists as they did in the past. The sad message is that
ordinary citizens who use YouTube are not able to recollect the missing verification after a few days. The well known *sleeper effect* (Hovland/Janis/Kelley 1953)
tells us that after a few days, people will remember the core message of the video
but not the source. As we are forced to watch more and more unverified material
via professional news outlets, news consumers should evaluate the journalistic
standards of *news tips*.

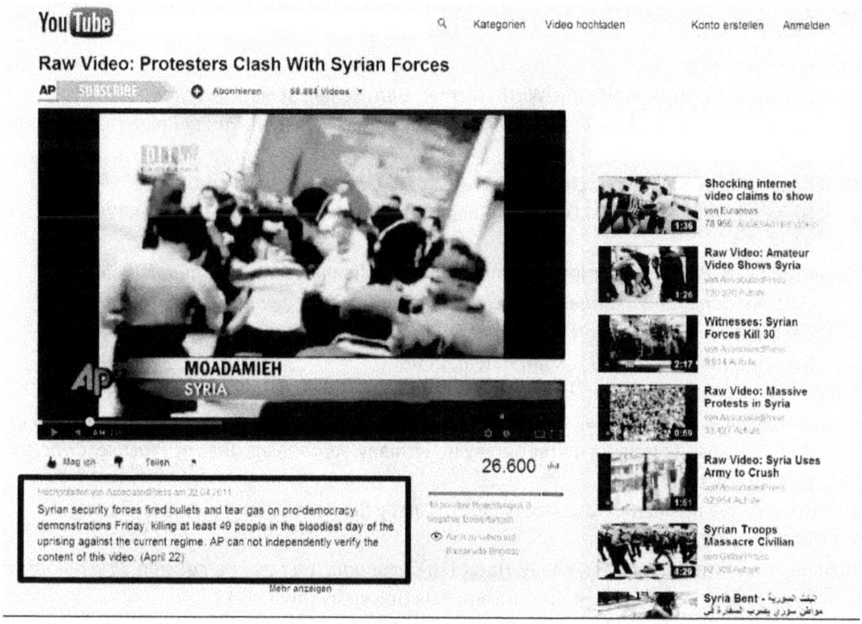

Figure 4: Screenshot of an *AP*-released video on YouTube without proper verification.

3 Conclusion

User participation is rapidly evolving. In particular, more and more professional media outlets make use of the integration of user-generated content in their products. The question whether stand-alone citizen journalism products will permanently reach big audiences and will provide proper quality at the same time is still open. But it seems plausible that the benefits of swarm intelligence and crowdsourcing are limited to certain projects and that this form of journalism works well when there is a clearly defined task (like in the case of Wikipedia). Otherwise the costs of screening, selecting, and verifying news might easily outweigh the benefits. Originally *pure* citizen journalism projects, like the South Korean website *OhmyNews,* used professional editors in order to deal with the vast amount of information received. In doing so, professional and citizen journalism converge as citizen journalism becomes more and more professionalized and re-institutionalized.

References

Beck, Klaus/Schweiger, Wolfgang/Wirth, Werner. *Gute Seiten, schlechte Seiten: Qualität in der Onlinekommunikation*. Internet Research. Bd. 15. München: Verlag Reinhard Fischer, 2004.

Busemann, Katrin/Gscheidle, Christoph. Web 2.0: Aktive Mitwirkung verbleibt auf niedrigem Niveau. Ergebnisse der ARD/ZDF-Onlinestudie 2011. *Media Perspektiven* 7–8 (2011): 360–369.

Engesser, Sven. Partizipativer Journalismus: Eine Begriffsanalyse. In *Kommunikation, Partizipation und Wirkungen im Social Web. Band 2 Strategien und Anwendungen: Perspektiven für Wirtschaft Politik und Publizistik*, Zerfass, Ansgar/Welker, Martin/Schmidt, Jan (eds.), 47–71. Köln: Halem, 2008.

Fröhlich, Romy/Quiring, Oliver/Engesser, Sven. Between idiosyncratic self-interests and professional standards: a contribution to the understanding of participatory journalism in Web 2.0. Results from an online survey in Germany. *Journalism: Theory, Practice, Criticism* (2012 forthcoming).

Gillmor, Dan. *We the Media: Grassroots Journalism by the People, for the People*. Beijing: O'Reilly, 2006.

Hovland, Carl I./Janis, Irving L./Kelley, Harold H. *Communication and Persuasion. Psychological Studies of Opinion Change*. New Haven: Yale University Press, 1953.

Howe, Jeff. The rise of crowdsourcing. Retrieved on 17th January 2012, from http://www.wired.com/wired/archive/14.06/crowds.html (2006).

McCombs, Maxwell E. *Setting the Agenda: The Mass Media and Public Opinion* (Reprint). Cambridge: Polity Press, 2008.

Meier, Klaus. *Journalistik*. 2nd ed. Konstanz: UVK-Verlagsgesellschaft, 2011.

Neuberger, Christoph/Nuernbergk, Christian. Competition, complementarity or integration? The relationship between professional and participatory media. *Journalism Practice* 4, no. 3 (2010): 319–332.

Paulussen, Steve/Domingo, David/Heinonen, Ari/Singer, Jane/Quandt, Thorsten/Vujnovic, Marina. Citizen participation in online news media. An overview of current developments in four European countries and the United States. In *Journalismus online – Partizipation oder Profession?* Quandt, Thorsten/Schweiger, Wolfgang (eds.), 263–283. Wiesbaden: VS Verlag für Sozialwissenschaften/GWV Fachverlage GmbH Wiesbaden, 2008.

Purcell, Kristen/Rainie, Lee/Mitchell, Amy/Rosenstiel, Tom/Olmstead, Kenny. Understanding the participatory news consumer. How Internet and cell phone users have turned news into a social experience. Pew Research Center. Retrieved on 17th January 2012, from http://www.pewinternet.org/~/media//Files/Reports/2010/PIP_Understanding_the_Participatory_News_Consumer.pdf (2010).

Rogers, Everett M. *Diffusion of Innovations*. 5th ed. New York: Free Press, 2003.

Schmidt, Jan. *Weblogs: Eine kommunikationssoziologische Studie*. Konstanz: UVK-Verlagsgesellschaft, 2006.

Schmidt, Jan. *Das neue Netz: Merkmale, Praktiken und Folgen des Web 2.0*. Konstanz: UVK-Verlagsgesellschaft, 2009.

Schweiger, Wolfgang/Quiring, Oliver. User-generated content on mass media web sites – just a variety of interactivity or something completely different? Paper presented at the 55th

Annual Conference of the International Communication Association, New York, U.S., May 26th–30th, 2005.

Ziegele, Marc/Quiring, Oliver. In search for an online discussion value – assessing media-initiated user communication from a news value perspective: Paper presented at the 61st Annual Conference of the International Communication Association, Boston, U.S., May 26th–30th, 2011.

Part 5: **Conclusions**

Stephan Russ-Mohl
Journalism's neglected self-inspection

Final remarks

---

1 The uncovered story of the century

Providing the final conclusions for such a book is certainly a risky challenge. It is risky because everything has already been said, though not by everybody. The final contributor should certainly try hard to not become the everybody who says everything once again. I will therefore experiment with an old journalistic recipe called the "KISS" approach, designed to catch the attention once more. "KISS" meaning, Keep It Short and Simple.

A few final thoughts on why journalism is and will remain important for our societies and democracies, and why we should continue to think about journalism's future:

In her most recent book, Anya Schiffrin (Columbia University) and a handful of experts – among them Nobel Prize-winning economist Joseph Stiglitz – have analyzed how "America's business press missed the story of the century" (Schiffrin 2011). They indicate how a lack of resources, pack behavior, and increasing stress in newsrooms are undermining professional standards and journalistic quality. For several years the Pew Research Center's Project for Excellence in Journalism (2004 ff.), research from the Reuters Institute for the Study of Journalism in Oxford (Levy/Nielsen 2010; Lloyd 2011), and the annual yearbook "Qualität der Medien" in Switzerland (fög 2010, 2011), have – among many others (Meyer 2004; Russ-Mohl 2009; Kramp/Weichert 2012) – arrived at similar conclusions.

Stiglitz states that, "a critical press might serve as one of the checks and balances, restoring sanity to markets that have lost touch with reality" (Stiglitz 2011: 23). But he is also fully aware of why such high expectations are difficult to meet. "Reporters and their editors do not stand apart from the rest of society. They too can easily be swept up in the herd mentality," because "there are strong incentives for the media not to lean against the wind" (Stiglitz 2011: 24). He identifies a significant problem in the symbiotic relationship between the press and those they cover. This cozy relationship "does not necessarily serve the rest of the society well." Moreover, "hubris can lead to the view by journalists that as recipients of information they can sort out the distortions and inaccuracies – so long as they can get the information," (Stiglitz 2011: 26) – a viewpoint which has been reflected more recently by Stefan Geiss (2012) as "Besserwisser-Journalismus" ("know better journalism").

2 The neglected public interest

Let us enlarge the perspective by considering a few of 2011's other major events, besides the rescue umbrellas for those European countries which, by all standards of accounting, have seemed to be bankrupt for quite a while. At a superficial first glance, topics like the Ehec virus, the nuclear meltdown in Fukushima, the abrupt turn-around in energy politics, and even the bloody massacre of Anders Breivik who terrorized Norway, have little in common – except that they were breathtaking events for all of us, and that the media successfully captured our attention in covering them.

However upon second glance, these topics have more in common. Let us consider for a moment what would have happened if media outlets and journalists had served the public interest properly in covering these events. One might argue that these events would either not have taken place at all, or that they would have developed quite differently.

Though the public interest may be difficult to define (Downs 1962; Whittle/ Cooper 2009), it is obvious that both commercial and public media outlets are neglecting it in their day-to-day operations. They frequently deliver one-sided and sensationalized news reports – or, in economists' terms, demeritory rather than meritory goods. Unfortunately, many newsrooms are even cheating their publics by transforming PR into journalism, without further checks and without contributing any added value.

Had the journalists investigated before ringing the Ehec alarm, the Ehec media circus would not have taken place. It cost farmers and governments millions of euros – according to Haller (2011), solely German taxpayers were charged 200 million euros for the destructed crops.

Had the media not looked the other way for so many years, the energy turn in Germany and Switzerland would have taken place less abruptly and with less absurd results. Even today nobody knows how radioactive material can be disposed of safely. If the media in France and in the Czech Republic had provided less lickspittle journalism, it would probably be unthinkable that nuclear reactors would continue to operate there without any plans to close them down.

Even the terror attack in Norway is difficult to imagine, had the assassin not been able to predict that his desire for worldwide attention would be reliably satisfied by the media. Suggestions by researchers like Bruno Frey (2004) on how the media might react more responsibly to terrorist attacks have unfortunately never been taken seriously.

Had the press adopted its watchdog function, neither the last crash at the stock exchange nor the rescue umbrellas would have been necessary. If the early warning system of journalism had worked well, investment bankers couldn't have considered themselves as "masters of the universe," nor could governments in Europe have signed contracts and designed laws with which they obviously did not intend to comply. Perhaps we would never have had the euro, but certainly journalists would have provided a timely warning about the disaster facing Europe and the United States.

3 The twofold challenge for the future of journalism

According to Dean Starkman (2011) from the *Columbia Journalism Review*, alarm bells have been ringing. In a content analysis of the nine most influential business press outlets in the U.S., Starkman identified – in the time period between January 2000 and June 2007 – 730 articles containing significant warnings. However, this is not an impressive record. We should keep in mind that the *Wall Street Journal* alone published 220,000 stories during that period. According to Starkman, in a sense the 730 stories were "corks bobbing on a news Niagara."

Starkman's analysis provides empirical evidence that excellent and investigative reporting still can be found. But it also gives proof of how such analysis can be drowned by huge waves of PR. One-sided press releases flood newsrooms, and PR creates an undertow, which endangers journalistic credibility. One of the founding fathers of communications research, Emil Dovifat, was probably the first to observe in the early twentieth century that PR supports the laziness of reporters (Dovifat 1927: 209).

To summarize, the challenge for the future is twofold. The first and most urgent is finding a business model or a non-profit alternative to sustain high quality reporting in a convergent media world where advertising can bypass news, and where journalism is increasingly subsidized by PR.

The second challenge lies in keeping journalism secure from government, and creating more transparency with regard to the shortcomings of journalism. Such transparency might help newsrooms to maintain a balance which allows them to exercise at least some control over the PR spoon-feeding.

However, this may be wishful thinking. The most drastic cutbacks in newsrooms concern reporting about media and journalism. And the most recent debate in Switzerland about the annual "Quality of the Media" report presented by Kurt Imhof (fög 2010, 2011) clearly shows that journalists overreact if communication researchers make shortcomings of the journalism public. As competent media journalists are now absent in newsrooms, publishers and top editors themselves attacked the researchers as a kind of "last stand" – and thereby neglected the basic journalistic rules of balance, fairness, and disinterestedness (Supino 2010, Strehle 2010, Voigt 2011; Wälty 2011; Lebrument 2011, 2012).

This is not exactly the way media and journalism should deal with journalism research, with their own profession, and with the revolution of convergence. A strategy to regain the confidence of the audience and credibility of the media would look different.

References

Dovifat, Emil. *Der amerikanische Journalismus*. Stuttgart: Deutsche Verlags-Anstalt, 1927 (reprint: Russ-Mohl, Stephan (ed.). Berlin: Colloquium Verlag, 1990).

Downs, Anthony: The public interest: its meaning in a democracy. *Social Research* 29, no. 1 (1962): 1–36.

fög – Forschungsbereich Öffentlichkeit und Gesellschaft an der Universität Zürich (ed.). *Qualität der Medien: Schweiz – Suisse – Svizzera. Jahrbuch*. Basel: Schwabe Verlag, 2010.

fög – Forschungsbereich Öffentlichkeit und Gesellschaft an der Universität Zürich (ed.). *Qualität der Medien: Schweiz – Suisse – Svizzera. Jahrbuch*. Basel: Schwabe Verlag, 2011.

Frey, Bruno. *Dealing with Terrorism: Stick or Carrot?* Cheltenham: Elgar, 2004.

Geiss, Stefan. Medienberichterstattung über Journalismus- und Medienforschung. Presentation at the University of Mainz, 23rd June 2012.

Haller, Michael. Gurken, Keime, Kolportagen. *Message*, no. 3 (2011): 17.

Kramp, Leif/Weichert, Stephan. *Innovationsreport Journalismus. Ökonomische, medienpolitische und handwerkliche Faktoren im Wandel*. Bonn: Friedrich-Ebert-Stiftung, 2012.

Lebrument, Hanspeter. Dauerbrenner Qualität. In *Flash extra. Magazin des Verbandes Schweizer Medien*, Verband Schweizer Medien (ed.), 5. Zurich, 2011.

Lebrument, Hanspeter. Qualität: Wissenschaft ersetzt Auflage. In *Flash extra. Magazin des Verbandes Schweizer Medien*, Verband Schweizer Medien (ed.), 5. Zurich, 2012.

Levy, David A.L./Nielsen, Rasmus Kleis. *The Changing Business of Journalism and Its Implications for Democracy*. Oxford: Reuters Institute for the Study of Journalism, University of Oxford, 2010.

Lloyd, John: Scandal! News international and the rights of journalism. Retrieved on 21st August 2012, from http://reutersinstitute.politics.ox.ac.uk/publications/risj-challenges/scandal-news-international-and-the-rights-of-journalism.html (2011).

Meyer, Philip. *The Vanishing Newspaper: Saving Journalism in the Information Age*. Columbia, MO: University of Missouri Press, 2004.

Pew Research Center's Project for Excellence in Journalism. The state of the news media. Retrieved on 21st August 2012, from http://stateofthemedia.org/previous-reports (2004 ff.).

Russ-Mohl, Stephan. *Kreative Zerstörung. Niedergang und Neuerfindung des Zeitungsjournalismus in den USA*. Konstanz: UVK Verlagsgesellschaft, 2009.

Schiffrin, Anya (ed.). *Bad News. How America's Business Press Missed the Story of the Century*. New York/London: The New Press, 2011.

Starkman, Dean. Power problem. In *Bad News. How America's Business Press Missed the Story of the Century*, Schiffrin, Anya (ed.), 37–53. New York/London: The New Press, 2011.

Stiglitz, Joseph E. The media and the crisis. In *Bad News. How America's Business Press Missed the Story of the Century*, Schiffrin, Anya (ed.), 22–36. New York/London: The New Press, 2011.

Strehle, Res. Wie gut sind unsere Medien? Retrieved on 24th June 2011, from http://www.tagesanzeiger.ch/meinungen/dossier/kolumnen--kommentare/Wie-gut-sind-unsere-Medien/story/23903696 (2010).

Supino, Pietro. Die Qualität unserer Presse. Retrieved on 24th June 2011, from http://www.tagesanzeiger.ch/schweiz/standard/Die-Qualitaet-unserer-Presse/story/28385132/print.html (2010).

Voigt, Hansi. Falschaussage mit Qualitätsanspruch. Retrieved on 15th October 2011, from http://www.20min.ch/finance/news/story/Falschaussage-mit-Qualitaets-anspruch-20687990 (2011).

Wälty, Peter. Kritik der Kritik. Retrieved on 14th August 2012, from http://www.tagesanzeiger.ch/leben/gesellschaft/Kritik-der-Kritik/story/14198479 (2011).

Whittle, Stephen/Cooper, Glenda. Privacy, probity and public interest. Retrieved on 14th August 2012, from http://reutersinstitute.politics.ox.ac.uk/?id=455 (2009).

Bibliography

AFP. New York Times publishes "crowd-funded" article. Retrieved on 13th November 2009, from http://www.google.com/hostednews/afp/article/ALeqM5iSC_k8BSFHlGN6n-bX0mpFbUaF1Tw (2009).

Anderson, Christopher W. Understanding the role played by algorithms and computational practices in the collection, evaluation, presentation, and dissemination of journalistic evidence. Conference draft. Paper prepared for the 1st Berlin Symposium on Internet and Society, 25th–27th October, 2011.

ARD-ZDF Onlinestudie. Retrieved on 10th September 2012, from http://www.ard-zdf-onlinestudie.de/ (2011).

Ariely, Dan. *Predictably Irrational. The Hidden Forces That Shape Our Decisions*. New York: HarperCollins, 2008, 2009.

Arnove, Robert/Pinede, Nadine. Revisiting the "big three" foundations. *Critical Sociology* 33 (2007): 389–425.

Arora, Neeraj/Dreze, Xavier/Ghose, Anindya/Hess, James D./Iyengar, Raghuram/Jing, Bing/Joshi, Yogesh/Kumar, V./Lurie, Nicholas/Neslin, Scott/Sajeesh, S./Su, Meng/Syam, Niladri/Thomas, Jacquelyn/Zhang, Z. John. Putting one-to-one-marketing to work: personalization, customization, and choice. *Marketing Letters* 19, no. 3/4 (2008): 305–321.

Asch, Solomon E. Effects of group pressure upon the modification and distortion of judgment. In *Groups, Leadership and Men*, Guetzkow, Harold S. (ed.), 177–190. Pittsburgh: Carnegie Press, 1951.

Baerns, Barbara. Vielfalt und Vervielfältigung. Befunde aus der Region – eine Herausforderung für die Praxis. *Media Perspektiven* 3 (1983): 207–215.

Baerns, Barbara. Macht der Öffentlichkeitsarbeit und Macht der Medien. In *Politikvermittlung. Beiträge zur politischen Kommunikationskultur. Schriftenreihe der Bundeszentrale für politische Bildung (238)*, Sarcinelli, Ulrich (ed.), 147–160. Bonn: Verlag Bonn Aktuell, 1987.

Baerns, Barbara. *Öffentlichkeitsarbeit oder Journalismus? Zum Einfluss im Mediensystem*. Revised ed. (1st ed. 1985). Köln: Verlag Wissenschaft und Politik, 1991.

Baerns, Barbara. The "Determination Thesis": How independent is journalism of public relations? In *A Complicated, Antagonistic & Symbiotic Affair. Journalism, Public Relations and their Struggle for Public Attention*, Merkel, Bernd/Russ-Mohl, Stephan/Zavaritt, Giovanni (eds.), 43–57. Lugano: Università della Svizzera Italiana, Giampiero Casagrande editore, 2007.

Baerns, Barbara. Public relations is what public relations does. Conclusions from a long-term project on professional public relations. In *Public Relations Metrics. Research and Evaluation*, Van Ruler, Betteke/Tkalac Vercic, Ana/Vercic, Dejan (eds.), 154–169. New York, London: Routledge, 2008.

Baerns, Barbara. Öffentlichkeitsarbeit und Erkenntnisinteressen der Publizistik- und Kommunikationswissenschaft. In *Theorien der Public Relations. Grundlagen und Perspektiven der PR-Forschung*, Röttger, Ulrike (ed.), 285–297. 2nd revised ed. Wiesbaden: VS Verlag für Sozialwissenschaften, 2009.

Bakardjieva, Maria. *Internet Society: The Internet in Everyday Life*. London: Sage, 2005.

Beck, Klaus/Schweiger, Wolfgang/Wirth, Werner. *Gute Seiten, schlechte Seiten: Qualität in der Onlinekommunikation*. Internet Research. Bd. 15. München: Verlag Reinhard Fischer, 2004.

Beck, Klaus/Dogruel, Leyla/Reineck, Dennis. Wirtschaft Online: Zweitverwertung oder publizistischer Mehrwert? Eine Analyse aus Kommunikatorsicht. *Publizistik* 55 (2010): 231–251.

Bentele, Günter/Grosskurth, Lars/Seidenglanz, René. *Profession Pressesprecher: Vermessung eines Berufsstandes*. Berlin: Helios Media, 2005, 2007, 2009.

Berman, Saul J./Battino, Bill/Shipnuck, Louisa/Neus, Andreas. The end of advertising as we know it. IBM Institute for Business Value. Retrieved on 21st November 2007, from http://www-935.ibm.com/services/us/gbs/thoughtleadership/?cntxt=a1000062 (2007).

Berman, Saul J./Battino, Bill/Feldman, Karen. Beyond content. Capitalizing on the new revenue opportunities. IBM Institute for Business Value. Retrieved on 22nd August 2010, from http://public.dhe.ibm.com/common/ssi/ecm/en/gbe03361usen/GBE03361USEN.PDF (2010).

Bernays, Edward L. *Crystallizing Public Opinion*. New York: Kessinger Publishing, 2004.

Bernays, Edward L. *Propaganda*. New York: Ig Publishing, 2005.

Boston Consulting Group (BCG). Press release – news for sale: charges for online news are set to become the norm as most consumers say they are willing to pay, according to The Boston Consulting Group. Retrieved on 10th September 2012, from http://www.bcg.com/media/PressReleaseDetails.aspx?id=tcm:12-35297 (2009).

Bowman, Shayne/Willis, Chris. We media. How audiences are shaping the future of news and information. *NDN*. Retrieved on 12th November 2011, from http://www.hypergene.net/wemedia/download/we_media.pdf (2003).

Bruce, Ian R. Public relations and advertising. In *The Advertising Business. Operations, Creativity, Media Planning, Integrated Communications*, Jones, John Philip (ed.), 473–483. London/New Delhi: Thousand Oaks, 1999.

Bundesverband Deutscher Zeitungsverleger e.V. (BDZV). *Jahrbuch Zeitungen 2007*. Berlin: ZV Zeitungs-Verlag Service GmbH, 2007.

Bundesverband Deutscher Zeitungsverleger e.V. (BDZV). *Jahrbuch Zeitungen 2010*. Berlin: ZV Zeitungs-Verlag Service GmbH, 2010.

Burda, Hubert. Conference: DLD (Digital – Life – Design). Munich, 2009.

Busemann, Katrin/Gscheidle, Christoph. Web 2.0: Aktive Mitwirkung verbleibt auf niedrigem Niveau. Ergebnisse der ARD/ZDF-Onlinestudie 2011. *Media Perspektiven* 7–8 (2011): 360–369.

Campbell, Margaret C. Perceptions of price unfairness: antecedents and consequences. *Journal of Marketing Research* 36, no. 2 (1999): 187–199.

CBS News. US Funded Arab TV's credibility crisis. 60 Minutes. Retrieved on 12th May 2009, from http://www.cbsnews.com/stories/2008/06/19/60minutes/main4196477.shtml (2008).

Centre For Public Inquiry. Press release 16th December 2005. Retrieved on 7th August 2009, from http://www.publicinquiry.ie/pressreleases.php (2005).

Centre For Public Inquiry. Statement of Frank Connolly. 7th December 2005. Retrieved on 7th August 2009, from http://www.publicinquiry.ie/pressreleases.php (2005).

Centre For Public Inquiry. The great corrib gas controversy. Fiosrú 1, no. 2, 2005.

Chang, Byeng-Hee/Chan-Olmsted, Sylvia M. Relative constancy of advertising spending. A crossnational examination of advertising expenditures and their determinants. *Gazette, The International Journal for Communication Studies* 67, no. 4 (2005): 339–357.

Cialdini, Robert B. *Influence: The Psychology of Persuasion*. New York: HarperCollins, 1998.

Collins, Lauren. House perfect. Is the IKEA ethos comfy or creepy? *The New Yorker*, 3rd October 2011: 54–59.

Com.X/prmagazin. Entspannte Verhältnisse? *Prmagazin* 41, no. 12 (2010): 34–39.

Currah, Andrew. *What's Happening to Our News: An Investigation into the Likely Impact of the Digital Revolution on the Economics of News Publishing in the UK.* Oxford: University of Oxford, Reuters Institute for the Study of Journalism, 2009.

Davies, Nick. *Flat Earth News.* London: Chatto & Windus, 2008.

Deleersnyder, Barbara/Dekimpe, Marnik G./Steenkamp, Jan-Benedict E.M./Leeflang, Peter S.H. The role of national culture in advertising's sensitivity to business cycles: An investigation across continents. *Journal of Marketing Research* 46, no. 5 (2009): 623–636.

Deutscher Bundestag. Zwischenbericht der Bundesregierung über die Lage von Presse und Rundfunk in der Bundesrepublik Deutschland. Drucksache 6/692, 1970.

Deutscher Bundestag. Unterrichtung durch den Beauftragten der Bundesregierung für Kultur und Medien. Medien- und Kommunikationsbericht der Bundesregierung. Drucksache 16/11570, 2008.

Dewenter, Ralf. Die Preissetzung von Internet Content Providern: Oder naht das Ende der „Gratiskultur"? Zeitgespräch: Was darf das Internet kosten? *Wirtschaftsdienst* 89, no. 10 (2009): 657–659.

DIPR (Deutsches Institut für Public Relations). Primärerhebung. Berufsbild Public Relations in der BRD. Köln: Unpublished Manuscript, 1973.

Dobelli, Rolf. *Die Kunst des klaren Denkens.* München: Hanser, 2011.

Dovifat, Emil. *Der amerikanische Journalismus.* Stuttgart: Deutsche Verlags-Anstalt, 1927 (reprint: Russ-Mohl, Stephan (ed.). Berlin: Colloquium Verlag, 1990).

Downs, Anthony: The public interest: its meaning in a democracy. *Social Research* 29, no. 1 (1962): 1–36.

Edmonds, Rick. Getting behind the media: What are the subtle trade-offs for foundation-funded journalism? Retrieved on 12th May 2009, from http://www.philanthropyroundtable.org/article.asp?article=1104&paper=1&cat=147 (2002).

Engesser, Sven. Partizipativer Journalismus: Eine Begriffsanalyse. In *Kommunikation, Partizipation und Wirkungen im Social Web. Band 2 Strategien und Anwendungen: Perspektiven für Wirtschaft Politik und Publizistik,* Zerfass, Ansgar/Welker, Martin/Schmidt, Jan (eds.), 47–71. Köln: Halem, 2008.

Erichsen, Jens. The new media ecosystem. Unpublished manuscript for the conference "Media Convergence & Journalism", University of Mainz, 21st–22nd October, 2011.

Fasenfest, David. Notes from the editor. *Critical Sociology* 33 (2007): 381–382.

Feldman, Bob. Report from the field: left media and left think tanks – foundation-managed Protest? *Critical Sociology* 33 (2007): 427–446.

Fielding, Nick/Cobain, Ian. Revealed: US spy operation that manipulates social media. Retrieved on 12th August 2012, from http://www.guardian.co.uk/technology/2011/mar/17/us-spy-operation-social-networks (2011).

Firger, Jessica. Protesters' newspaper occupies a familiar name. Retrieved on 14th May 2012, from http://blogs.wsj.com/metropolis/2011/10/04/protesters-newspaper-occupies-a-familiar-name (2011).

Foa, Marcello. *Gli stregoni della notizia.* Milano: Guerini e Associati, 2006.

fög – Forschungsbereich Öffentlichkeit und Gesellschaft an der Universität Zürich (ed.). *Qualität der Medien: Schweiz – Suisse – Svizzera. Jahrbuch.* Basel: Schwabe Verlag, 2010.

fög – Forschungsbereich Öffentlichkeit und Gesellschaft an der Universität Zürich (ed.). *Qualität der Medien: Schweiz – Suisse – Svizzera. Jahrbuch.* Basel: Schwabe Verlag, 2011.

Frey, Bruno. *Dealing with Terrorism: Stick or Carrot?* Cheltenham: Elgar, 2004.

Fröhlich, Romy/Quiring, Oliver/Engesser, Sven. Between idiosyncratic self-interests and professional standards: a contribution to the understanding of participatory journalism in Web 2.0. Results from an online survey in Germany. *Journalism: Theory, Practice, Criticism* (2012 forthcoming).

Geiss, Stefan. Medienberichterstattung über Journalismus- und Medienforschung. Presentation at the University of Mainz, 23rd June 2012.

GfK-Verein. Die Bereitschaft, für Internetinhalte zu bezahlen, ist gering – Internationale GfK-Studie zur Internetnutzung in 17 Ländern. Retrieved on 13th September 2012, from http://www.gfk.com/imperia/md/content/presse/091211_wsje_internet_dfin.pdf (2009).

Gieseking, Thomas. *Gewinnoptimale Preisbestimmung in werbefinanzierten Märkten. Eine conjoint-analytische Untersuchung eines Publikumszeitschriftenmarktes.* Wiesbaden: Gabler, 2010.

Gillmor, Dan. *We the Media: Grassroots Journalism by the People, for the People.* Beijing: O'Reilly, 2006.

Goldmark, Peter. President's letter. In annual report of the Rockefeller Foundation. Retrieved on 6th August 2009, from www.rockfound.org/library/annual_reports/1990–1999/1997.pdf (1997).

Grimes, Arthur/Rae, David/O'Donovan, Brendan. Determinants of advertising expenditures: Aggregate and cross-media evidence. *International Journal of Advertising* 19, no. 3 (2000): 317–334.

Guensburg, Carol. Nonprofit news: as news organizations continue to cut back, investigative and enterprise journalism funded by foundations and the like is coming to the fore. *American Journalism Review* 30, no. 1 (2008): 26–33.

Guilhot, Nicholas. Reforming the world: George Soros, global capitalism and the philanthropic management of the social sciences. *Critical Sociology* 33 (2007): 447–479.

Haller, Michael. Gurken, Keime, Kolportagen. *Message*, no. 3 (2011): 17.

Hallin, Daniel C. Commercialism and professionalism in the American news media. In *Mass Media and Society*, Curran, James/Gurevitch, Michael (eds.), 218–237. London: Arnold, 2000.

Hallin, Daniel C./Mancini, Paolo. *Comparing Media Systems: Three Models of Media and Politics.* Cambridge: Cambridge University Press, 2005.

Hamilton Consultants/Deighton, John/Quelch, John. Economic value of the advertising-supported Internet ecosystem. Retrieved on 10th September 2012, from http://www.iab.net/media/file/Economic-Value-Report.pdf (2009).

Harvey, David. Organizing for the anti-capitalist transition. Talk for the World Social Forum 2010, Porto Alegre, Brazil. Retrieved on 18th March 2010, from http://davidharvey.org/2009/12/organizing-for-the-anti-capitalist-transition/#more-376 (2009).

Hoffrage, Ulrich. Overconfidence. In *Cognitive Illusions: A Handbook on Fallacies and Biases in Thinking, Judgement and Memory*, Pohl, Rüdiger F. (ed.), 235–254. Hove: Psychology Press, 2004.

Höhn, Tobias D. Schnittstelle Nachrichtenagenturen und Public Relations – eine Untersuchung am Beispiel der Deutschen Presse-Agentur (dpa). Unpublished Master's Thesis University of Leipzig, 2005.

Holiday, Ryan. *Trust Me, I'm Lying: Confessions of a Media Manipulator.* New York: Portfolio/Penguin, 2012.

Hoshaw, Lindsey. A float in the Ocean. Expanding islands of trash. Retrieved on 13th
    November 2009, from http://www.nytimes.com/2009/11/10/science/10patch.html?_
    r=1&scp=1&sq=lindsey%20hoshaw&st=cse (2009).
Hovland, Carl I./Janis, Irving L./Kelley, Harold H. *Communication and Persuasion. Psychological
    Studies of Opinion Change.* New Haven: Yale University Press, 1953.
Howe, Jeff. The rise of crowdsourcing. Retrieved on 17th January 2012, from http://www.wired.
    com/wired/archive/14.06/crowds.html (2006).
IAB-PWC. IAB Internet advertising revenue report conducted by PricewaterhouseCoopers (PwC).
    Retrieved on 17th September 2009, from http://www.iab.net/insights_research/530422/
    adrevenuereport (2009).
Illich, Ivan/Zola, Irving Kenneth/McKnight, John/Caplan, Jonathan/Shaiken, Harley. *Disabling
    Professions.* New York/London: Marion Boyars, 2000.
Jarolimek, Stefan. *Die Transformation von Öffentlichkeit und Journalismus. Modellentwurf und
    das Fallbeispiel Belarus.* Wiesbaden: VS Verlag für Sozialwissenschaften, 2009.
Jones, Alex S. *Losing the News: The Future of News that Feeds Democracy.* Oxford: Oxford
    University Press, 2009.
Jones, Nicholas. *Sultans of Spin.* London: Orion, 2000.
Jones, Nicholas. *The Control Freaks.* London: Politico's Publishing, 2002.
Kahneman, Daniel. *Thinking, Fast and Slow.* London: Penguin, 2011.
Kahneman, Daniel/Knetsch, Jack L./Thaler, Richard H. Fairness and the assumptions of
    economics. *Journal of Business 59*, no. 4 (1986): 285–300.
Kahneman, Daniel/Knetsch, Jack L./Thaler, Richard H. Fairness as a constraint on profit
    seeking: Entitlements in the market. *The American Economic Review 76*, no. 4 (1986):
    728–741.
Kahneman, Daniel/Knetsch, Jack L./Thaler, Richard H. Fairness as a constraint on profit
    seeking: Entitlements in the market. In *Choices, Values, and Frames*, Kahneman, Daniel/
    Tversky, Amos (eds.), 317–334. Cambridge: Cambridge University Press, 2000.
Katz, Elihu/Lazarsfeld, Paul F. *Personal Influence.* Glencoe, Il: Free Press, 1955.
Kaye, Jeff/Quinn, Stephen. *Funding Journalism in the Digital Age. Business Models, Strategies,
    Issues and Trends.* New York: Peter Lang, 2010.
Keen, Andrew. *The Cult of the Amateur: How Today's Internet is Killing Our Culture.* New York:
    Doubleday, 2007.
KEK (Kommission zur Ermittlung der Konzentration im Medienbereich). Press Release August
    2009. Ausschreibung der KEK für ein Gutachten zum Thema „Die Bedeutung des Internets
    im Rahmen der Vielfaltssicherung". Retrieved on 19th August 2009, from www.kek-online.
    de (2009).
Kerl, Katharina. Das Bild der Public Relations in der Berichterstattung ausgewählter deutscher
    Printmedien. Eine quantitative Inhaltsanalyse. Unpublished Master's Thesis University of
    Munich, 2007.
Klaue, Siegfried/Knoche, Manfred/Zerdick, Axel (eds.). *Probleme der Pressekonzentrations-
    forschung. Ein Experten-Colloquium an der Freien Universität Berlin. Materialien zur
    interdisziplinären Medienforschung 12.* Baden-Baden: Nomos, 1980.
Kopp, Sven/Nienstedt, Heinz-Werner. Wie wichtig ist wahrgenommene Preisfairness für die
    Akzeptanz der Einführung kostenpflichtiger journalistischer Nachrichtenangebote im
    Internet? Working Paper Medienmanagement Mainz, 2–2012.
Kramp, Leif/Weichert, Stephan. *Innovationsreport Journalismus. Ökonomische, medienpo-
    litische und handwerkliche Faktoren im Wandel.* Bonn: Friedrich-Ebert-Stiftung, 2012.

Küng, Lucy/Picard, Robert G./Towse, Ruth. *The Internet and the Mass Media*. London: Sage, 2008.

Lamey, Lien/Deleersnyder, Barbara/Dekimpe, Marnik G./Steenkamp, Jan-Benedict E.M. How business cycles contribute to private-label success: Evidence from the U.S. and Europe. *Journal of Marketing* 71, no. 1 (2007): 1–15.

Larson, Rob. The Clinton Foundation Donors. Retrieved on 18th March 2010, from http://www.counterpunch.org/larson01282009.html (2009).

Lazarsfeld, Paul F./Berelson, Bernard/Gaudet, Hazel. *The People's Choice. How the Voter Makes Up His Mind in a Presidential Campaign*. New York: Columbia University Press, 1948.

Lebrument, Hanspeter. Dauerbrenner Qualität. In *Flash extra. Magazin des Verbandes Schweizer Medien*, Verband Schweizer Medien (ed.), 5. Zurich, 2011.

Lebrument, Hanspeter. Qualität: Wissenschaft ersetzt Auflage. In *Flash extra. Magazin des Verbandes Schweizer Medien*, Verband Schweizer Medien (ed.), 5. Zurich, 2012.

Levy, David A.L./Nielsen, Rasmus Kleis. *The Changing Business of Journalism and Its Implications for Democracy*. Oxford: Reuters Institute for the Study of Journalism, University of Oxford, 2010.

Linnett, Richard. Magazines pay the price of TV recovery. *Advertising Age* 73, no. 35 (2002): 1–2.

Lischka, Juliane/Kienzler, Stephanie/Siegert, Gabriele. Advertising expenditures, consumer spending and business expectations. Assessing these relations for German data from 1991–2009. Presentation accepted for the European Communication Research and Education Association (ECREA) Conference 2012, "Social media and global voices", 24th–27th October 2012, Istanbul, TR, 2012.

Lloyd, John: Scandal! News international and the rights of journalism. Retrieved on 21st August 2012, from http://reutersinstitute.politics.ox.ac.uk/publications/risj-challenges/scandal-news-international-and-the-rights-of-journalism.html (2011).

Löffelholz, Martin. Dimensionen struktureller Kopplung von Öffentlichkeitsarbeit und Journalismus. Überlegungen zur Theorie selbstreferentieller Systeme und Ergebnisse einer repräsentativen Studie. In *Aktuelle Entstehung von Öffentlichkeit. Akteure – Strukturen – Veränderungen,* Bentele, Günter/Haller, Michael (eds.), 187–208. Konstanz: UVK, 1997.

Machill, Marcel/Beiler, Markus/Zenker, Martin unter Mitarbeit von Gerstner, Johannes R. *Journalistische Recherche im Internet. Bestandsaufnahme journalistischer Arbeitsweisen in Zeitungen, Hörfunk, Fernsehen und Online. Schriftenreihe Medienforschung der Landesanstalt für Medien Nordrhein-Westfalen 60*. Berlin: Vistas, 2008.

Machill, Marcel/Beiler, Markus/Gerstner, Johannes R. *Online-Recherchestrategien für Journalistinnen und Journalisten. Workshopmaterialien für die Aus- und Weiterbildung*. Edited by Landesanstalt für Medien Nordrhein-Westfalen. Wuppertal: Boerje Halm, 2009.

Maletzke, Gerhard. *Psychologie der Massenkommunikation*. Hamburg: Hans Bredow-Institut, 1963.

Markoff, John. Apple's visionary redefined digital age. Retrieved on 14th May 2012, from http://www.nytimes.com/2011/10/06/business/steve-jobs-of-apple-dies-at-56.html?_r=1&pagewanted=all (2011).

Mazzoleni, Gianpietro. *La comunicazione politica*. Bologna: Il Mulino, 2004.

McChesney, Robert W./Nichols, John. *The Death and Life of American Journalism: The Media Revolution That Will Begin the World Again*. Philadelphia: Nation Books, 2010.

McCombs, Maxwell E. *Setting the Agenda: The Mass Media and Public Opinion* (Reprint). Cambridge: Polity Press, 2008.

Meckel, Miriam. *Next: Erinnerungen an eine Zukunft ohne uns*. Reinbek: Rowohlt, 2011.

Meier, Klaus. *Journalistik*. 2nd ed. Konstanz: UVK-Verlagsgesellschaft, 2011.

Meier, Klaus/Reimer, Julius. Transparenz im Journalismus. Instrumente, Konfliktpotentiale, Wirkung. *Publizistik* 56 (2011): 133–155.

Meyer, Philip. *The Vanishing Newspaper: Saving Journalism in the Information Age*. Columbia, MO: University of Missouri Press, 2004.

Michel Kommission. Bericht der Kommission zur Untersuchung der Wettbewerbsgleichheit von Presse, Funk/Fernsehen und Film. Bundestags-Drucksache 5/2120, 28th September 1967.

Miner, Michael. Is Pro Publica living up to its promise? Retrieved on 14th May 2009, from http://www.chicagoreader.com/TheBlog/archives/2008/07/07/pro-publica-living-its-promise (2008).

Mintz, Anne P. (ed.). *Web of Deceit: Misinformation and Manipulation in the Age of Social Media*. Medford: Information Today, 2012.

Neuberger, Christoph. Alles Content, oder was? Vom Unsichtbarwerden des Journalismus im Internet. In *Innovationen im Journalismus: Forschung für die Praxis*, Hohlfeld, Ralf/Meier, Klaus/Neuberger, Christoph (eds.), 25–69. Münster: Lit, 2002.

Neuberger, Christoph/Nuernbergk, Christian/Rischke, Melani. Journalismus im Internet: Zwischen Profession, Partizipation und Technik. *Media Perspektiven* 47 (2009): 174–188.

Neuberger, Christoph/Lobigs, Frank. *Die Bedeutung des Internets im Rahmen der Vielfalts-sicherung*. Gutachten im Auftrag der Kommission zur Ermittlung der Konzentration im Medienbereich (KEK). Berlin: Vistas, 2010.

Neuberger, Christoph/Nuernbergk, Christian. Competition, complementarity or integration? The relationship between professional and participatory media. *Journalism Practice* 4, no. 3 (2010): 319–332.

Neuberger, Christoph/vom Hofe, Hanna Jo/Nuernbergk, Christian. Twitter und Journalismus. Der Einfluss des „Social Web" auf die Nachrichten. LfM-Dokumentation 38. 3rd ed. Düsseldorf: Landesanstalt für Medien Nordrhein-Westfalen (LfM). Retrieved on 3rd November 2011, from http://lfmpublikationen.lfm-nrw.de/catalog/downloadproducts/L043_Band_38_Twitter.pdf (2011).

Newspaper Association of America (NAA). Trends and numbers. Retrieved on 13th September 2012, from http://www.naa.org/Trends-and-Numbers.aspx (2012).

Nienstedt, Heinz-Werner/Ebel, Fridtjof. Pricing für App und Online: Eine conjointanalytische Untersuchung überregionaler Nachrichtenangebote. Working Paper Medienmanagement Mainz, 1–2012.

Nocera, Joe. Self-made philanthropists. Retrieved on 8th December 2009, from http://www.nytimes.com/2008/03/09/magazine/09Sandlers-t.html (2008).

Noelle-Neumann, Elisabeth. Wirkung der Massenmedien. In *Publizistik*, Noelle-Neumann, Elisabeth/Schulz, Winfried (eds.), 316–350. Frankfurt am Main: S. Fischer, 1971.

Nothhaft, Howard. Rezension zu Ulrike Röttger, Joachim Preusse und Jana Schmitt: Grundlagen der Public Relations. Eine kommunikationswissenschaftliche Einführung. *Publizistik* 57 (2012): 253–254.

O'Clery, Conor. *The Billionaire Who Wasn't: How Chuck Feeney Secretly Made and Gave Away a Fortune*. New York: Public Affairs, 2007.

Olmstead, Kenny/Mitchell, Amy/Rosenstiel, Tom. Navigating news online: Where people go, how they get there and what lures them away. Retrieved on 12th November 2011, from http://www.journalism.org/sites/journalism.org/files/NIELSEN%20STUDY%20-%20Copy.pdf (2011).

Onlinevermarkterkreis im Bundesverband Digitale Wirtschaft e.V. (OVK). Online-Werbeinvestitionen nähern sich 2011 der 6-Milliarden-Euro-Grenze. Retrieved on 13th September 2012, from http://www.ovk.de/ovk/ovk-de/online-werbung/datenfakten/ werbeinvestitionen-nach-segmenten.html (2012).

Paulussen, Steve/Domingo, David/Heinonen, Ari/Singer, Jane/Quandt, Thorsten/Vujnovic, Marina. Citizen participation in online news media. An overview of current developments in four European countries and the United States. In *Journalismus online – Partizipation oder Profession?* Quandt, Thorsten/Schweiger, Wolfgang (eds.), 263–283. Wiesbaden: VS Verlag für Sozialwissenschaften/GWV Fachverlage GmbH Wiesbaden, 2008.

Perez-Latre, Francisco J. The paradigm shift in advertising and its meaning for advertising-supported media. *Journal of Media Business Studies* 4, no. 18 (2007): 41–49.

Perry, Suzanne. Financier backs project to beef up investigative reporting. Retrieved on 15th May 2009, from http://philanthropy.com/free/articles/v20/i02/02001001.htm (2007).

Pew Research Center's Project for Excellence in Journalism. The state of the news media. An annual report on American journalism. Retrieved on 13th September 2012, from http://stateofthemedia.org/ (2004 ff.).

Picard, Robert G. Effects of recessions on advertising expenditures: An exploratory study of economic downturns in nine developed nations. *Journal of Media Economics* 14, no. 1 (2001): 1–14.

Picard, Robert G. *Evolution of Revenue Streams and the Business Model of Newspapers: The U.S. Industry between 1950–2000*, Discussion Papers C1/2002, Business Research and Development Centre, Turku School of Economics and Business Administration, 2002.

Picard, Robert G. Capital crisis in the profitable newspaper industry. *Nieman Reports* 60, no. 4 (2006): 10–12.

Picard, Robert G. Shifts in newspaper advertising expenditures and their implications for the future of newspapers. *Journalism Studies* 9, no. 5 (2008): 704–716.

Picard, Robert G. The future of the news industry. In *Media and Society*, Curran, James (ed.), 365–379. London: Bloomsbury Academic, 2010.

Picard, Robert G. *The Economics and Financing of Media Companies*. 2nd ed. New York: Fordham University Press, 2011.

Pickard, Victor/Stearns, Josh/Aaron, Craig. *Saving the News: Toward a National Journalism Strategy.* Washington, D.C.: Free Press, 2009.

Picot, Arnold. Erlöspolitik für Informationsangebote im Internet. Zeitgespräch: Was darf das Internet kosten? *Wirtschaftsdienst* 89, no. 10 (2009): 643–647.

Plotkowiak, Thomas/Stanoevska-Slabeva, Katarina/Ebermann, Jana/Meckel, Miriam/Fleck, Matthes. Netzwerk-Journalismus. Zur veränderten Vermittlerrolle von Journalisten am Beispiel einer Case Study zu Twitter und den Unruhen in Iran. *Medien & Kommunikationswissenschaft* 60 (2012): 102–124.

PR-Trendmonitor. PR-Trend 2012. *PR Report* 12 (2011): 11.

PricewaterhouseCoopers (PwC). *Global entertainment and media outlook: 2008–2012, Industry Overview, 9th annual edition.* New York, 2008.

PricewaterhouseCoopers (PwC). *Global entertainment and media outlook: 2010–2014, Industry Overview, 11th annual edition.* New York, 2010.

Purcell, Kristen/Rainie, Lee/Mitchell, Amy/Rosenstiel, Tom/Olmstead, Kenny. Understanding the participatory news consumer. How Internet and cell phone users have turned news into a social experience. Retrieved on 12th November 2011, from http://www.journalism.org/sites/journalism.org/files/Participatory_News_Consumer.pdf (2010).

Rainey, James. Los Angeles Times receives $1-million grant from Ford Foundation. Retrieved on 10th September 2012, from http://www.editorandpublisher.com/Newsletter/Article/Los-Angeles-Times-Receives--1-Million-Grant-From-Ford-Foundation (2012).

Reilly Center for Media and Public Affairs. *The Breaux Symposium: New Models for News*. Baton Rouge: Louisiana State University, 2008.

Rochet, Jean-Charles/Tirole, Jean. Platform competition in two-sided markets. *Journal of the European Economic Association* 1, no. 4 (2003): 990–1029.

Rochet, Jean-Charles/Tirole, Jean. Two-sided markets: A progress report. *RAND Journal of Economics* 37, no. 3 (2006): 645–667.

Roelofs, David. Note on this special issue of Critical Sociology. *Critical Sociology* 33 (2007): 387–388.

Rogers, Everett M. *Diffusion of Innovations*. 5th ed. New York: Free Press, 2003.

Rolke, Lothar. Journalisten und PR-Manager – eine antagonistische Partnerschaft mit offener Zukunft. In *Wie die Medien Wirklichkeit steuern und selbst gesteuert werden*, Rolke, Lothar/Wolff, Volker (eds.), 223–247. Wiesbaden: Westdeutscher Verlag, 1999.

Ronneberger, Franz. *Kommunikationspolitik I. Institutionen, Prozesse, Ziele*. Kommunikations-wissenschaftliche Bibliothek 6. Mainz: v. Hase & Koehler, 1978.

RTE. Secret Billionaire: The Chuck Feeney story (TV documentary broadcast on 5th May 2009).

Ruch, Karl-Heinz. Personal talk with Stephan Russ-Mohl, 2nd May in Berlin, 2012.

Russ-Mohl, Stephan. The economics of journalism and the challenge to improve journalism quality. A research manifesto. *Studies in Communication Sciences 6*, no. 2 (2006): 189–208.

Russ-Mohl, Stephan. *Kreative Zerstörung. Niedergang und Neuerfindung des Zeitungsjour-nalismus in den USA*. Konstanz: UVK Verlagsgesellschaft, 2009.

Russ-Mohl, Stephan. Opfer der Medienkonvergenz? Wissenschaftskommunikation und Wissen-schaftsjournalismus im Internet-Zeitalter. In *Medienkonvergenz – Transdisziplinär*, Füssel, Stephan (ed.), 81–108. Berlin/Boston: De Gruyter, 2012.

Saksena, Shashank/Hollifield, C. Ann. U.S. newspapers and the development of online editions. *International Journal on Media Management* 4, no. 2 (2002): 75–84.

Schiffrin, Anya (ed.). *Bad News. How America's Business Press Missed the Story of the Century*. New York/London: The New Press, 2011.

Schmidt, Holger. Trafficlieferanten der Medien: Facebook gewinnt, Google verliert. *faz. net*. Netzökonom. Retrieved on 12th November 2011, from http://faz-community.faz. net/104216/print.aspx (2011).

Schmidt, Jan. *Weblogs: Eine kommunikationssoziologische Studie*. Konstanz: UVK-Verlagsge-sellschaft, 2006.

Schmidt, Jan. *Das neue Netz: Merkmale, Praktiken und Folgen des Web 2.0*. Konstanz: UVK-Verlagsgesellschaft, 2009.

Scholl, Armin/Weischenberg, Siegfried. *Journalismus in der Gesellschaft. Theorie, Methodologie und Empirie*. Wiesbaden: Westdeutscher Verlag, 1998.

Schultz, Don E. Integrated marketing communications and how it relates to traditional media advertising. In *The Advertising Business. Operations, Creativity, Media Planning, Integrated Communications*, Jones, John Philip (ed.), 325–338. London/New Delhi: Thousand Oaks, 1999.

Schweiger, Wolfgang/Quiring, Oliver. User-generated content on mass media web sites – just a variety of interactivity or something completely different? Paper presented at the 55th

Annual Conference of the International Communication Association, New York, U.S., May 26th–30th, 2005.

Selbach, David. Lauschangriff. *Prmagazin* 42, no. 10 (2011): 60–65.

Shafer, Jack. Nonprofit journalism comes at a cost. Retrieved on 13th November 2009, from http://www.slate.com/id/2231009/ (2009).

Shaver, Mary A./Shaver, Dan. Changes in the levels of advertising expenditures during recessionary periods: A study of advertising performance in eight countries. Paper presented at the Asian-American Academy of Advertising, Hong Kong, 2005.

Siebert, Fred S./Peterson, Theodore/Schramm, Wilbur. *Four Theories of the Press. The Authoritarian, Libertarian, Social Responsibility and Soviet Communist Concepts of What the Press Should Be and Do.* Urbana, Illinois: University of Illinois Press, 1956.

Siegert, Gabriele. Online-Kommunikation und Werbung. In *Handbuch Online-Kommunikation*, Schweiger, Wolfgang/Beck, Klaus (eds.), 434–460. Wiesbaden: VS Verlag für Sozialwissenschaften, 2010.

Siegert, Gabriele/Brecheis, Dieter. *Werbung in der Medien- und Informationsgesellschaft. Eine kommunikationswissenschaftliche Einführung.* Wiesbaden: VS Verlag für Sozialwissenschaften, 2005.

Siegert, Gabriele/Trepte, Sabine/Baumann, Eva/Hautzinger, Nina. Qualität gesundheitsbezogener Online-Angebote aus Sicht von Usern und Experten. *Medien & Kommunikationswissenschaft* 53 (2005): 486–506.

Siegert, Gabriele/Kienzler, Stephanie/Lischka, Juliane/Mellmann, Ulrike. Medien im Sog des Werbewandels. Konjunkturell und strukturell bedingte Veränderungen der Werbeinvestitionen und Werbeformate und ihre Folgen für die Medien. Report. (Project funded by the Swiss National Foundation), 2012.

Sklair, Leslie. Achilles has two heels: Crises of capitalist globalization. In *Thinker, Faker, Spinner, Spy: Corporate PR and the Assault on Democracy*, Dinan, William/Miller, David (eds.), 21–32. London: Pluto Press, 2007.

Southwell, Brian G./Yzer, Marco C. The roles of interpersonal communication in mass media campaigns. In *Communication Yearbook 31*, Beck, Christina (ed.), 420–462. New York: Routledge, 2007.

Southwell, Brian G./Yzer, Marco C. (eds.). Conversation and Campaigns. Communication Theory 19, no. 1 (2009): 1–101.

Stannard-Stockton, Sean. Philanthropists' "soft power" may trump the hard pull of purse strings. Retrieved on 22nd April 2010, from http://philanthropy.com/article/Soft-Power-Could-Be-More/65080/ (2010).

Starkman, Dean. Power problem. In *Bad News. How America's Business Press Missed the Story of the Century*, Schiffrin, Anya (ed.), 37–53. New York/London: The New Press, 2011.

Staun, Harald. Was genau war denn früher besser? In *Frankfurter Allgemeine Sonntagszeitung*, 29th July 2012.

Stiglitz, Joseph E. The media and the crisis. In *Bad News. How America's Business Press Missed the Story of the Century*, Schiffrin, Anya (ed.), 22–36. New York/London: The New Press, 2011.

Strehle, Res. Wie gut sind unsere Medien? Retrieved on 24th June 2011, from http://www.tagesanzeiger.ch/meinungen/dossier/kolumnen--kommentare/Wie-gut-sind-unsere-Medien/story/23903696 (2010).

Stroehlein, Andrew. PAX: A new idea in conflict prevention? Retrieved on 14th May 2012, from http://www.andrewstroehlein.com/2010/05/pax-new-idea-in-conflict-prevention.html (2010).

Strom, Stephanie. To advance their cause, foundations buy stocks. Retrieved on 2nd December 2011, from http://www.nytimes.com/2011/11/25/business/foundations-come-to-the-aid-of-companies.html?_r=1&emc=tnt&tntemail0=y (2011).

Supino, Pietro. Die Qualität unserer Presse. Retrieved on 24th June 2011, from http://www.tagesanzeiger.ch/schweiz/standard/Die-Qualitaet-unserer-Presse/story/28385132/print.html (2010).

Surowiecki, James. *The Wisdom of Crowds*. New York: Anchor Books, 2005.

Szyszka, Peter/Schütte, Dagmar/Urbahn, Katharina. *Public Relations in Deutschland. Eine empirische Studie zum Berufsfeld Öffentlichkeitsarbeit*. Konstanz: UVK, 2009.

Thier, Dave. How this guy lied his way into *MSNBC, ABC News, The New York Times* and more. Retrieved on 12th August 2012, from http://www.forbes.com/sites/davidthier/2012/07/18/how-this-guy-lied-his-way-into-msnbc-abc-news-the-new-york-times-and-more (2012).

Time. 25 people to blame for the financial crisis. Retrieved on 14th May 2009, from http://www.time.com/time/specials/packages/article/0,28804,1877351_1877350_1877343,00.html (2009).

Trappel, Josef/Nieminen, Hannu/Nord, Lars (eds.). *The Media For Democracy Monitor. A Cross National Study of Leading News Media*. University of Gothenburg: Nordicom, 2011.

Turner, Sebastian. How advertising changes in the digital revolution and how this affects news media. Unpublished manuscript for the conference "Media Convergence & Journalism", University of Mainz, 21st–22nd October, 2011.

Vallone, Robert P./Griffin, Dale W./Lin, Sabrina/Ross, Lee. Overconfident prediction of future actions and outcomes by self and others. *Journal of Personality and Social Psychology* 58, no. 4 (1990): 582–592.

van der Wurff, Richard/Bakker, Piet/Picard, Robert G. Economic growth and advertising expenditures in different media in different countries. *Journal of Media Economics* 21, no. 3 (2008): 28–52.

van Dijk, Tuen A. News, discourse and ideology. In *The Handbook of Journalism Studies*, Wahl-Jorgensen, Karin/Hanitzsch, Thomas (eds.), 191–204. London: Routledge, 2009.

Verband Deutscher Zeitschriftenverleger (VDZ). *New Media Trends & Insights*. Berlin, 2012.

Voigt, Hansi. Falschaussage mit Qualitätsanspruch. Retrieved on 15th October 2011, from http://www.20min.ch/finance/news/story/Falschaussage-mit-Qualitaetsanspruch-20687990 (2011).

Wälty, Peter. Kritik der Kritik. Retrieved on 14th August 2012, from http://www.tagesanzeiger.ch/leben/gesellschaft/Kritik-der-Kritik/story/14198479 (2011).

Wan-Ifra. *World Press Trends 2010 Edition*. Darmstadt, 2010.

Weischenberg, Siegfried/Malik, Maja/Scholl, Armin. *Die Souffleure der Mediengesellschaft. Report über die Journalisten in Deutschland*. Konstanz: UVK Verlagsgesellschaft, 2006.

Weischenberg, Siegfried/Malik, Maja/Scholl, Armin. Journalismus in Deutschland 2005. Zentrale Befunde der aktuellen Repräsentativbefragung deutscher Journalisten. *Media Perspektiven* 44 (2006): 346–361.

Westphal, David. Philanthropic foundations: Growing funders of the news. USC Annenberg School for communication. Center on communication leadership & Policy research series.

Retrieved on 7th August 2009, from http://communicationleadershipblog.uscannenberg.
   org/Westphal-Philanthropic%20Support%20for%20News%20report.pdf (2009).

Whittle, Stephen/Cooper, Glenda. Privacy, probity and public interest. Retrieved on 14th August
   2012, from http://reutersinstitute.politics.ox.ac.uk/?id=455 (2009).

Wiencierz, Christian. An expert's view on conversations and advertising. Presentation at the
   Advertising in Communication and Media Research Symposium 14th–15th June 2012,
   University of Tuebingen, 2012.

Wirtz, Bernd. *Medien- und Internetmanagement*, 2nd ed. Wiesbaden: Gabler, 2001.

Wladarsch, Jennifer. Auf der Spur der Massenmedien in sozialen Onlinenetzwerken. Wie und
   warum Internetnutzer massenmediale Inhalte in sozialen Onlinenetzwerken nutzen.
   Unpublished Master's Thesis Ludwig-Maximilians-University Munich, 2010.

World Internet Project. World Internet Project report finds large percentages of non-users, and
   significant gender disparities in going online. Retrieved on 10th September 2012, from
   http://www.digitalcenter.org/WIP2010/wip2010_long_press_release_v2.pdf (2010).

Yzer, Marco C./Southwell, Brian G. New communication technologies, old questions. *American
   Behavioral Scientist* 52, no. 1 (2008): 8–20.

Zak, Elana. The Guardian shares newslists with readers. Retrieved on 14th May 2012, from
   http://www.mediabistro.com/10000words/new-experiment-lets-readers-influence-
   editorial-decision-making-process-at-the-guardian_b7513 (2011).

Zeitungs Marketing Gesellschaft mbH & Co. KG (ZMG). Standortbestimmung Junge Leser:
   Desktop-Research – Eine erste Grundlagenstudie zum Thema im Auftrag des BDZV.
   Frankfurt am Main, 2011.

Zeller, Frauke/Wolling, Jens. Struktur und Qualitätsanalyse publizistischer Onlineangebote.
   Überlegungen zur Konzeption der Online-Inhaltsanalyse. *Media Perspektiven* 48 (2010):
   143–153.

Zentralverband der Deutschen Werbewirtschaft (ZAW). Medien: Die meisten im Plus. Retrieved
   on 13th September 2012, from http://www.zaw.de/index.php?menuid=119 (2012).

Ziegele, Marc/Quiring, Oliver. In search for an online discussion value – assessing media-
   initiated user communication from a news value perspective: Paper presented at the 61st
   Annual Conference of the International Communication Association, Boston, U.S., May
   26th–30th, 2011.

Authors and editors

Dr. Barbara Baerns is a professor emerita of the theory and practice of journalism and public relations at the Freie Universität in Berlin, Germany. She spent several years teaching media and communication science as a professor at the Ruhr University in Bochum, Germany (1982–1989), in addition to holding positions in political journalism and public relations.

Harry Browne is a lecturer of journalism at the Dublin Institute of Technology, Ireland, where he teaches journalism practice, press history, and political communication. Furthermore, he is an author, a contributor to many academic and journalistic publications, and a frequent guest on Irish radio and TV programs. He is pursuing a Ph.D. at the National University of Ireland.

Marcello Foa is CEO of *Timedia*, a publishing group based in Lugano, Switzerland. Furthermore, he is a lecturer at the Università della Svizzera italiana in Lugano, cofounder of the European Journalism Observatory (EJO), and an author. His main research field is spin doctoring.

Konny Gellenbeck is director of the *tageszeitung (taz)* cooperative in Berlin, Germany. She has been working for the *taz* since 1986 and is one of the initiators of the *Panter Stiftung* in Berlin, a non-profit organization that, among other activities, supports young journalists.

Dr. Klaus Kocks is a journalist, columnist, and managing partner at the CATO Sozietät für Kommunikationsberatung GmbH in Horbach, Germany. He was engaged in consulting activities for the energy and automobile industry. Furthermore, he holds the title of honorary professor of Strategic Communication Management at the University of Applied Sciences in Osnabrück.

Peter Laufer, Ph.D., holds the James Wallace Chair in Journalism at the University of Oregon School of Journalism and Communications, U.S. He is an author and an award winning multi-platform journalist. A longtime *NBC News* correspondent, he has covered wars, natural disasters and social changes worldwide, including the collapse of the Berlin Wall.

Dr. Bettina Lis is an assistant professor at the Chair for Media Management at the Johannes Gutenberg University Mainz, Germany, since 2010. Before that, she was research assistant at the Department of Business Administration at the Johannes Gutenberg University Mainz and completed her Ph.D. in 2010. She holds a double degree in Communication Sciences (M.A.) and Business Administration (Diploma). Lectureships and research stays led her to the Warsaw School of Economics (Poland), the Dongbei University of Finance and Economics (PR China), and the University of Michigan, Ann Arbor (U.S.). Her main research interests are media management, marketing, and corporate communications.

Dr. Joachim Meinhold is CEO of the *Saarbrücker Zeitung Publishing Group*. After initially studying law and economics, he wrote his dissertation in the field of law before launching a professional career in the steel industry and the venture capital business. In 1992 he entered the media industry were he served as director of law and acquisitions for the *Publishing Group Georg von Holtzbrinck*. In 1993 he became CEO of the Holtzbrinck subsidiary *Der Tagesspiegel*, a Berlin based newspaper publishing house, before eventually becoming CEO of the *Saarbrücker Zeitung Publishing Group* in 2008.

Dr. Christoph Neuberger is a professor at the Department of Communication Science and Media Research (Institut für Kommunikationswissenschaft und Medienforschung, IfKW) at the Ludwig Maximilians University in Munich, Germany. His research focuses on the Internet (public sphere, online journalism, activities of the press and broadcasting media on the Internet, search engines, participatory formats) and journalism (theory, quality of news production, labor market, education).

Dr. Heinz-Werner Nienstedt is a professor at the Department of Communication (Institut für Publizistik) at the Johannes Gutenberg University Mainz, Germany, since 2002. Before that, he was CEO of the *Handelsblatt Publishing Group* and at the same time member of the Executive Board of the *Publishing Group Georg von Holtzbrinck*. His main research fields are media management and marketing. He is also an advisor to media firms and investors in these fields.

Dr. Robert G. Picard is a professor and director of research at the Reuters Institute for the Study of Journalism at the University of Oxford, U.K. He is the author and editor of 27 books and has been editor of the *Journal of Media Economics* and the *Journal of Media Business Studies*. He has consulted and carried out assignments for governments, international organizations, and leading media companies worldwide.

Dr. Oliver Quiring is a professor at and director of the Department of Communication (Institut für Publizistik) at the Johannes Gutenberg University Mainz, Germany. Furthermore, he is deputy chairman of the German Society for Journalism and Communication Science (Deutsche Gesellschaft für Publizistik- und Kommunikationswissenschaft, DGPuK). His main research fields are social, economic, and political communication, media change and media innovation, and qualitative and quantitative research methods.

Dr. Stephan Russ-Mohl is a professor of journalism and media management at the Università della Svizzera italiana in Lugano, Switzerland, director of the European Journalism Observatory (EJO), and at present a Gutenberg Fellow of the Research Unit Media Convergence at the University of Mainz, Germany. His main research fields are the economics of journalism, quality management in newsrooms, media and science journalism, and comparative journalism research.

Dr. Gabriele Siegert is a professor at and director of the Institute of Mass Communication and Media Research at the University of Zurich, Switzerland (Institut für Publizistikwissenschaft und Medienforschung der Universität Zürich, IPMZ). Her main research fields are media economics, media management, and advertising.

Bartosz Wilczek is a Ph.D. student at the Institute for Media and Journalism at the Università della Svizzera italiana, Switzerland, and at present a research assistant at the Research Unit Media Convergence at the Johannes Gutenberg University Mainz, Germany. His main research interests are the economics of journalism, newsroom convergence, quality assurance, and political and science journalism.

www.ingramcontent.com/pod-product-compliance
Lightning Source LLC
Chambersburg PA
CBHW070040100426
42740CB00013B/2744